ALGORITHMS FOR COMPUTER-AIDED DESIGN OF LINEAR MICROWAVE CIRCUITS

For a complete list of *The Artech House Microwave Library*, turn to the back of this book. . .

ALGORITHMS FOR COMPUTER-AIDED DESIGN OF LINEAR MICROWAVE CIRCUITS

Stanisław Rosłoniec

Artech House
Boston • London

Library of Congress Cataloging-in-Publication Data

Rosloniec, Stanislaw.
[Algorytmy projektowania wybranych liniowych ukladow mikrofalowych. English]
Algorithms for computer-aided design of linear microwave circuits / Stanislaw Rosloniec.
p. cm.
Translation of: Algorytmy projektowania wybranych liniowych ukladow mikrofalowych.
Includes bibliographical references and index.
ISBN 0-89006-354-0
1. Microwave integrated circuits—Design and construction—Data processing. 2. Computer-aided design. I. Title.
TK7876.R67 1990 90-42482
621.381'32—dc20 CIP

685 Canton Street
Norwood, MA 02062

International Standard Book Number: 0-89006-354-0
Library of Congress Catalog Card Number: 90-42482

10 9 8 7 6 5 4 3 2 1

CONTENTS

PREFACE

This book concerns microwave circuits with emphasis on linear passive ones composed of *RLC* lumped elements and commensurate TEM (or quasi-TEM) transmission line segments. As a rule, the design of this type of circuit requires specialized numerical techniques, and therefore a computer-aided design (CAD) approach is followed throughout the book. In the case of integrated microwave circuits, such an approach is particularly helpful because their design and optimization are fairly complicated if conducted in any other way.

As a matter of fact, the book provides CAD algorithms for typical UHF and microwave circuits. The book's philosophy and organization are the following.

Chapter 1 presents some general considerations on the computer-aided design process. Emphasis is on an explanation of how a computer can be effectively used for interactive design.

Chapter 2 describes several general-purpose optimization methods that are also useful for design and optimization of microwave circuits. In the first section, a short description of constrained and unconstrained optimization problems is given. Next, the one-dimensional search techniques, such as direct sequential search, Fibonacci search technique, and search by the golden section, are discussed. Section 2.3 is concerned with some gradient optimization strategies. Consequently, the steepest descent, Davidon-Fletcher-Powell, and Fletcher-Reeves methods are presented with the corresponding flow diagrams and illustrative calculations. Penalty function methods are discussed in Section 2.4 in a similar manner.

From a practical standpoint, many even more difficult design problems can be converted into the equivalent optimization problems and then solved by using a suitably selected conventional technique. Here, some of the optimization methods presented have been implemented in computer programs of the MICRO design system, the full description of which is given in Chapter 3.

The main part of the MICRO system is Library L2, composed of the CAD algorithms described in Chapter 5. The algorithms make possible the design and optimization of commonly used microwave circuits such as broadband transmission line impedance transformers, branch and coupled transmission line directional couplers, three-port hybrid power dividers, resistance attenuators, narrow-band impedance matching circuits, and different types of reactance filters.

The design of basic single and coupled transmission lines is the subject of Chapter 6. Closed-form relations are given for the design of coaxial lines, slab lines, striplines, microstrip lines, their accuracies, and the ranges of their validity. In a similar manner, we describe coupled transmission structures, namely coupled slab lines, coupled triplate striplines, and coupled microstrip lines. The computational programs presented in this chapter form the library denoted as L3. In general, these programs are called directly by using their abbreviations given in Figure 3.2 and Table 3.3. Moreover, as shown in Figure 3.2, they are related to programs of Library L2 by a special control procedure and can be called indirectly as subroutines.

Some useful relationships between *ABCD*-, *Z*-, *Y*-, *T*-, and *S*-parameters of a two-port network are considered in Chapter 4. This chapter also describes the Smith chart impedance transformation and a rigorous algorithm for design of the generalized stepped transmission line. The computer programs related to these problems constitute the auxiliary library, L1.

In most of the computer programs presented, in addition to computational instructions, there are short procedures for protection from different kinds of errors. The essence of these *check and protection* instructions is explained in Chapter 3. To make these programs more reliable and to reduce the execution time as much as possible, the most effective mathematical methods have been selected for many of the problems to be solved. In most cases, the methods are applied successively when approaching a solution.

Our intention is to present all problems under discussion in clear and comprehensive form. Therefore, the problems being analyzed are given together with the mathematical essentials and references to the literature. Furthermore, the check calculations are given along with the corresponding computer programs written in BASIC. All of these programs are stored on the IBM PC floppy disk which is an optional supplement to this book.

To conclude, we have tried to present the subject of microwave circuit design so that it is readily understood and clearly applicable to make the present book more suitable for a large group of readers. The book should be particularly useful for microwave circuit designers and graduate students. However, the reader is to judge how well we have succeeded in this endeavor. Comments regarding the organization, clarity of presentation, usefulness of the book and any other helpful suggestion will be welcome.

Finally, I wish to thank Andrzej B. Przedpelski of A.R.F. Products, Inc. for his support and helpful suggestions.

STANISLAW ROSLONIEC
WARSAW, JANUARY 1990

Chapter 1
OUTLINE OF A DESIGN PROCESS

1.1 THE COMPUTER-AIDED DESIGN APPROACH

As has been confirmed in practice, most of the algorithms intended for the design of electrical circuits include some computational stages which must be repeated many times in a similar manner. Today these rather laborious calculations are largely done automatically by means of digital computers. Of significance from the standpoint of a designer is that these calculations are very quickly as well as reliably made by the computer. As a result, we are able to realize iterative modifications of the circuit components being designed to achieve the assumed frequency responses, for example.

The general purpose optimization methods developed for other disciplines are useful for this aim as well. In the literature (see, e.g., [1–5]), the design process realized in this way is known as *computer-aided design,* denoted as CAD.

The typical CAD process for microwave circuits is composed of the following stages.

1. Synthesis of the electrical scheme of the circuit being designed that satisfies all imposed requirements.
2. Optimization of the circuit synthesized in Stage 1 with respect to its important parameter of interest, frequency response.
3. Realization of this circuit by calculating its structural parameters, mainly geometrical dimensions. The flow diagram for such a typical design procedure is shown in Figure 1.1.

According to the scheme presented above, the design process starts with a given set of specified electrical requirements. Let us assume further that the initial circuit configuration is selected on the basis of available design data and earlier experience. The responses of interest for this initial circuit are evaluated by means of suitable computer-aided analysis programs. Therefore, the numerical models of the various components of the circuit being analyzed must be prepared in a form suitable for CAD. As a rule, these models are contained in the analysis programs,

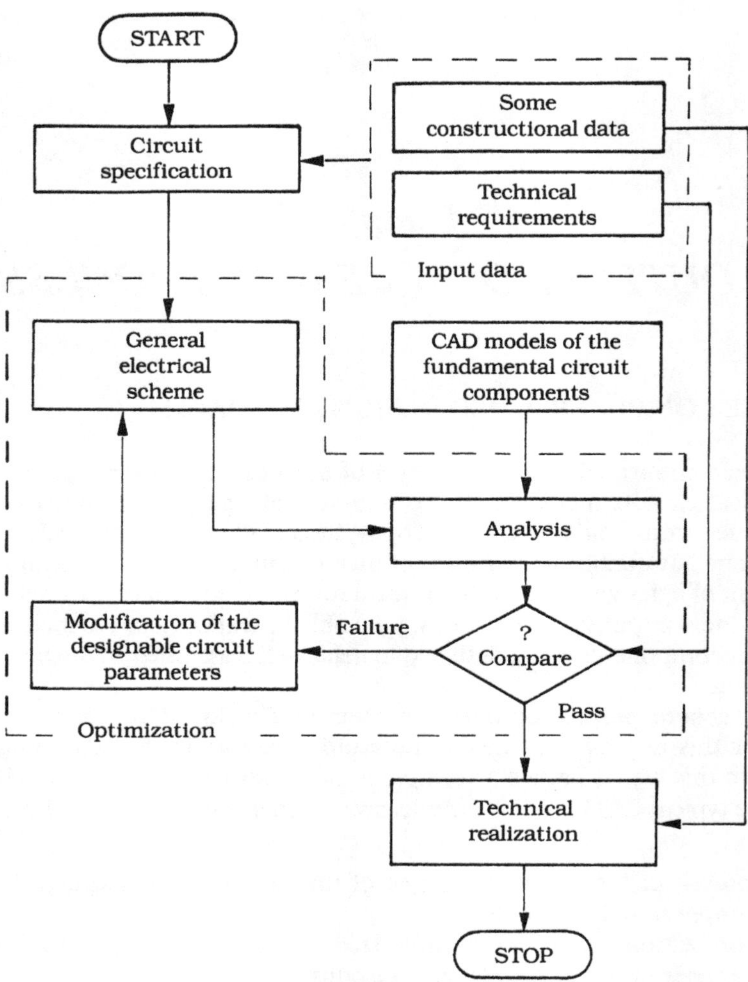

Figure 1.1 Conventional CAD procedure for microwave circuits.

which are often called *subroutines.* The given electrical requirements are compared with the corresponding circuit responses obtained as a result of analysis. If these results fail to satisfy the desired specifications, the design parameters of the circuit are varied according to a pre-set optimization strategy.

The sequence of circuit analysis, comparison of the results obtained with the desired performance, and the modification of the circuit parameters are continued iteratively until all the requirements are satisfied or the optimum circuit response is achieved under the given design constraints.

The final circuit configuration thus obtained is used later as the basis for construction of the real-world device. The computer is also helpful at this stage, especially when the circuit being designed includes some segments of different types of single and coupled transmission lines. In most cases, the design of these lines for given values of the characteristic impedances and other structural parameters consists of determining their geometrical dimensions. From the mathematical point of view, this is equivalent to solving the corresponding simultaneous nonlinear equations. In Section 2.1 and Chapter 6, we show how these equations can be successfully solved by using the appropriate minimization methods.

In the computer-aided design approach presented above, we assume that the electrical configuration of the circuit being designed is given first. Unfortunately, such an assumption causes a loss of generality of the design process. Therefore, in practice, different initial circuit configurations should be considered and compared. Undoubtedly, this can be done for only a few, practical configurations with numerical models stored in the computer memory. Moreover, the comparison criterion utilized should be formulated in a manner that ensures the best *univocal* choice. Under this criterion, not only the electrical specifications, but also the assumed structural constraints should be taken into account.

In general, the main advantage of the computer-aided design approach is that easily accessible general-purpose computers can be used. If the models of the fundamental circuit components are accurate enough, interactive design of many more complicated circuits is possible. Of course, the experimental verification of these circuits is necessary only at the final design stage. Hence, computer-aided design, in addition to its useful features, also has an interesting didactic aspect. In strict interpretation, this design can be viewed as a discourse between the author of the program and a designer. The computer acts immediately (and relatively quickly) to show computational results as well as appropriate comments. This is particularly important when inappropriate commands or incorrect input data are used. In this way, even more complicated design algorithms can be made accessible to many users with very small risk of an error being introduced.

REFERENCES

[1] Kuo, F.F., and W.G. Magnuson, *Computer Oriented Circuit Design,* Englewood Cliffs, NJ: Prentice-Hall, 1969.

[2] Calahan, D.A., *Computer Aided Network Design,* New York: McGraw-Hill, 1972.

[3] Chua, L.O., and P.M. Lin, *Computer Aided Analysis of Electronic Circuits: Algorithms and Computational Techniques,* Englewood Cliffs, NJ: Prentice-Hall, 1975.

[4] Gupta, K.C., R. Garg, and R. Chadha, *Computer-Aided Design of Microwave Circuits,* Norwood, MA: Artech House, 1981.

[5] Pozar, D., *Antenna Design Using Personal Computers,* Norwood, MA: Artech House, 1985.

Chapter 2
SOME OPTIMIZATION METHODS

2.1 GENERAL DESCRIPTION OF AN OPTIMIZATION PROBLEM

Many of the design algorithms considered in this book have been formulated as minimization problems (see Sections 5.11 and 6.5). Therefore, here we discuss some basic concepts and computational algorithms of optimization. In practice, optimization problems are most frequently presented as minimization of a scalar objective function $f(\bar{x})$, where $\bar{x} = (x_1, x_2, \ldots, x_n)$ are design parameters. Such an approach does not cause any loss of generality because the minima of function $f(\bar{x})$ correspond to the maxima of function $-f(\bar{x})$ [1–4]. Here, we note that the objective function $f(\bar{x})$ is also called the *error function* because, in most practical cases, it represents a difference between the achieved and desired performances. Thus, by a proper choice of the objective function, any optimization problem may be converted into the corresponding one of minimization.

In general, a constrained minimization problem consists of minimizing the given objective function $f(\bar{x})$. For this purpose, the values of the variables $\bar{x} = (x_1, x_2, \ldots, x_n)$ are modified during the minimization process in an appropriate manner. These variables, however, must satisfy the constraints given by the inequalities:

$$
\begin{aligned}
&r_1(x_1, x_2, \ldots, x_n) \leqslant 0 \\
&r_2(x_1, x_2, \ldots, x_n) \leqslant 0 \\
&\ldots \\
&r_k(x_1, x_2, \ldots, x_n) \leqslant 0
\end{aligned}
\tag{2.1}
$$

and equalities:

$$
\begin{aligned}
&h_{k+1}(x_1, x_2, \ldots, x_n) = 0 \\
&h_{k+2}(x_1, x_2, \ldots, x_n) = 0 \\
&\ldots \\
&h_{k+p}(x_1, x_2, \ldots, x_n) = 0
\end{aligned}
\tag{2.2}
$$

In most practical problems, equality constraints do not exist (i.e., conditions $h_j(x_1, x_2, \ldots, x_n) = 0$ for $j = k + 1, k + 2, \ldots, k + p$ do not exist).

The portion of the n-dimensional space of variables $\overline{x} = (x_1, x_2, \ldots, x_n)$ where all constraints are satisfied, is called the *feasible region* or the *design space,* and is usually denoted by R^n [3,5].

Now the minimization problem can be written in the following standard form:

$$\min_{\overline{x} \subset R^n} \quad f(\overline{x}) \tag{2.3}$$

where $\overline{x} \subset R^n$ means that we are seeking the minimum of function $f(\overline{x})$ inside the feasible region R^n.

For $R^n = E^n$, where E^n denotes the n-dimensional Euclidean space, (2.3) obviously reduces to its unconstrained version, namely

$$\min f(\overline{x}) \tag{2.4}$$

Simple examples of such unconstrained and constrained equations are given in Sections 2.32 and 2.41, respectively. Equation (2.4) is very much like finding a root. Many root-finding situations, if computationally convenient, can be formulated as minimization problems. Hence, let us consider, for example, k simultaneous equations:

$$\begin{aligned} \phi_1(x_1, x_2, \ldots, x_n) &= 0 \\ \phi_2(x_1, x_2, \ldots, x_n) &= 0 \\ &\ldots \\ \phi_k(x_1, x_2, \ldots, x_n) &= 0 \end{aligned} \tag{2.5}$$

The solution of this system of equations can be found by minimizing the function $U(\overline{x}) = U(x_1, x_2, \ldots, x_n)$, which is obtained by adding the squares of the left members $\phi_i(\overline{x})$. More strictly, because the solutions might be complex, the function to be minimized should be

$$U(\overline{x}) = \sum_{i=1}^{k} \phi_i(\overline{x})\phi_i^*(\overline{x}) \tag{2.6}$$

where $\phi_i^*(\overline{x})$ is the complex conjugate of $\phi_i(\overline{x})$. At the solution of the set of equations (2.5), simply we find that the function $U(\overline{x})$ achieves its minimum value equal to zero.

As has been confirmed experimentally, such an approach in seeking the roots is very useful in practice (see Chapter 6).

To solve (2.3) and (2.4), some of the optimization methods presented in Sections 2.2, 2.3, and 2.4 of this chapter can be used.

2.2 ONE-DIMENSIONAL MINIMIZATION TECHNIQUES

The one-dimensional search is widely applied for minimizing the given objective function of one or many variables. This search usually is to find the minimum value of the objective function in the direction determined by the variables of this function. This minimum value obviously can be achieved only if the derivative of the objective function is negative in the assumed search direction. As a rule, in one-dimensional minimization, the distance from the starting point is the auxiliary variable.

If the one-dimensional search is used for minimizing the multivariable objective function, the ith search iteration starts from point P_i of the feasible region in the selected ith direction. The starting point and direction are evaluated in the appropriate manner from the results of the previous iteration.

To find the minimum value of the objective function in the selected direction, we can use some elimination techniques presented in this section. Let us now assume that the objective function in the ith direction takes its minimum value at point Q_i. The search terminates if the obtained minimum is global. In other cases, the point Q_i serves as the starting point P_{i+1} for the next (i.e., $(i + 1)$th) iteration. Because the one-dimensional search is successively repeated, to develop efficient algorithms for this purpose is desirable. Three such conventional algorithms are presented below.

2.2.1 The Sequential Direct Search

In most minimization algorithms, the sequential direct search is first performed in order to find the feasible region (or an interval of the auxiliary variable) where the objective function is unimodal and takes its minimum value [3,4]. The essence of this simple elimination technique is illustrated in Figure 2.1 by the objective function $f(x_1, x_2)$ of two variables, denoted as x_1 and x_2. Let us assume that the starting point $\overline{x}^0 = [x_1^0, x_2^0]$ and direction $\overline{d} = [d_{x1}, d_{x2}]$ of search are given. In this case, variables x_1 and x_2 can be expressed in terms of auxiliary variable $t\overline{d}$, which is the distance from the starting point. In other words, $f(x_1, x_2)$ can be analyzed as the function of one variable, t. As a rule, in the search process, the variable t takes only discrete values $t_i = ih$, where h is the step and $i = 1, 2, \ldots$ The search of the interval containing the minimum terminates when $f(t_{i+1}) \geqslant f(t_i)$, and the interval $[t_i, t_{i+1}]$ is chosen for further analysis. For this purpose, the *Fibonacci search* technique or the *golden section* method can be used.

(a)

(b)

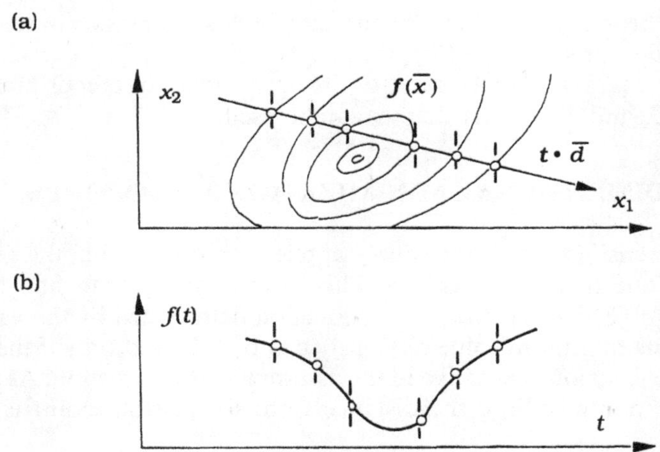

Figure 2.1 Typical level curves for a unimodal function of two variables.

2.2.2 The Fibonacci Search Technique

The search technique presented in this section derives its name from the Fibonacci sequence of numbers defined by

$$
\begin{aligned}
F_0 &= 1 \\
F_1 &= 1 \\
F_2 &= 2 \\
&\cdots \\
F_i &= F_{i-1} + F_{i-2}, \text{ for } i = 2, 3, 4, \ldots
\end{aligned}
\tag{2.7}
$$

Some first terms of this sequence are: 1, 1, 2, 3, 5, 8, 21, 34, 55, 89, . . .

The method presented starts with the assumption that the objective function $f(t)$ is unimodal in the interval $[a_0, b_0]$, where a_0 and b_0 are the lower and upper limits, respectively. At the beginning, the total number of iterations $n(n \geqslant 2)$ is evaluated from the following condition:

$$
|b_n - a_n| = |b_0 - a_0|/(2F_n) \tag{2.8}
$$

where a_n and b_n are the lower and upper limits of the nth searching interval.

Let us assume that after $k(k < n - 1)$ iterations, the searching interval reduces to $[a_k, b_k]$ (see Figure 2.2). The interior points t_k and t'_k of this interval are calculated as follows:

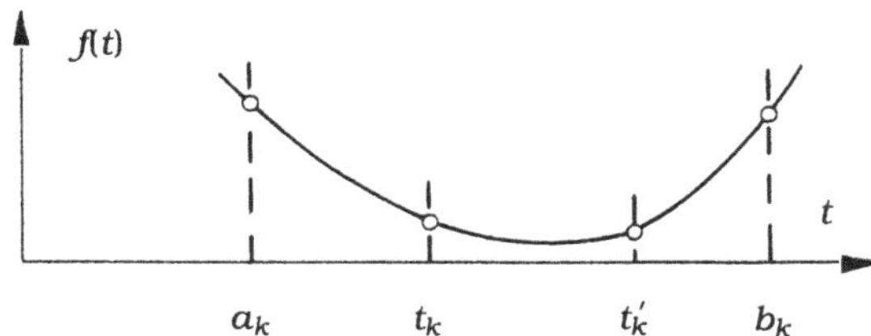

Figure 2.2 Profile of the linear search function $f(t)$.

$$t_k = a_k + (b_k - a_k)F_{n-1-k}/F_{n+1-k}$$
$$t'_k = a_k + (b_k - a_k)F_{n-k}/F_{n+1-k} \tag{2.9}$$

Let the values of the objective function evaluated at these points be $f(t_k)$ and $f(t'_k)$. There are three possibilities:

1. If $f(t_k) < f(t'_k)$, the minimum lies in the new $(k + 1)$-th interval $[a_{k+1}, b_{k+1}] = [a_k, t'_k]$;
2. If $f(t_k) > f(t'_k)$, the minimum lies in the interval $[a_{k+1}, b_{k+1}] = [t_k, b_k]$;
3. If $f(t_k) = f(t'_k)$, the intervals evaluated in examples 1 and 2 can be used as the $(k + 1)$-th interval because their lengths are equal.

In the last iteration, the points t_{n-1} and t'_{n-1} are calculated from

$$t_{n-1} = a_{n-1} + 0.5(b_{n-1} - a_{n-1})$$
$$t'_{n-1} = a_{n-1} + (0.5 + \epsilon)(b_{n-1} - a_{n-1}) \tag{2.10}$$

where ϵ is an optionally small positive number, for instance, $\epsilon = 10^{-6}$. If $f(t_{n-1}) < f(t'_{n-1})$, the minimum lies in the interval $[a_{n-1}, t_{n-1}]$. In other cases, the minimum being sought lies in the interval $[t_{n-1}, b_{n-1}]$. From the above considerations, the length of the last interval hence is equal ($|b_n - a_n| = 0.5\ |b_{n-1} - a_{n-1}|$), and is related to $|b_0 - a_0|$ by (2.8).

As has been confirmed experimentally, this search technique converges best for unimodal objective functions (see, e.g., [3,4]). In practice, however, its disadvantages are that the total number of iterations must be decided first, and all Fibonacci numbers must be stored in the computer memory.

2.2.3 Search by the Golden Section Technique

The one-dimensional search technique presented here is almost as effective as the Fibonacci search scheme described in the preceding section. This method has the advantage that the total number of iterations as well as function evaluations, n, need not be fixed in advance. Moreover, auxiliary number series need not be stored in the computer memory.

Let us consider again the problem of minimization of the unimodal objective function $f(t)$, such as shown in Figure 2.2. Now, the interior points t_k and t_k' of the kth interval are calculated from

$$t_k = a_k + (b_k - a_k)/s^2$$
$$t_k' = a_k + (b_k - a_k)/s \tag{2.11}$$

where $s = (1 + \sqrt{5})/2 = 1.618033\ldots$

In this case, the division of the interval yields two unequal parts such that the ratio of the whole to the larger section is equal to the ratio of the larger to the smaller section. This type of division of a linear segment is known as the *golden section* of a line [3].

The method presented consequently takes its name from this division. The choice of the next (i.e., $(k + 1)$th) interval is done in a manner similar to that employed in steps 1, 2, and 3 of the Fibonacci search technique (see Section 2.2.2). After n iterations, the length of the nth interval is found to be

$$|b_n - a_n| = |b_0 - a_0|/s^{n-1} \tag{2.12}$$

where a_0 and b_0 are the lower and upper limits of the beginning interval, respectively. The seeking of the minimum terminates if

$$|b_n - a_n| < \epsilon \tag{2.13}$$

where ϵ is an optionally small positive number. The flow diagram of this search technique is shown in Figure 2.3.

As has been confirmed in [3,4], search by the golden section method is only 17% slower than that using the Fibonacci technique. In practice, the present method is often preferred because the number of iterations need not be fixed in advance.

2.3 GRADIENT METHODS FOR MINIMIZATION

Let us consider a real function of n variables $f(\overline{x}) = f(x_1, x_2, \ldots, x_n)$, which is continuous and differentiable on E^n. The gradient of function $f(\overline{x})$ is a vector having components that are partial derivatives with respect to the particular variables:

$$\nabla f(\overline{x}) = \left[\frac{\partial f}{\partial x_1}, \frac{\partial f}{\partial x_2}, \ldots, \frac{\partial f}{\partial x_n}\right]^T \tag{2.14}$$

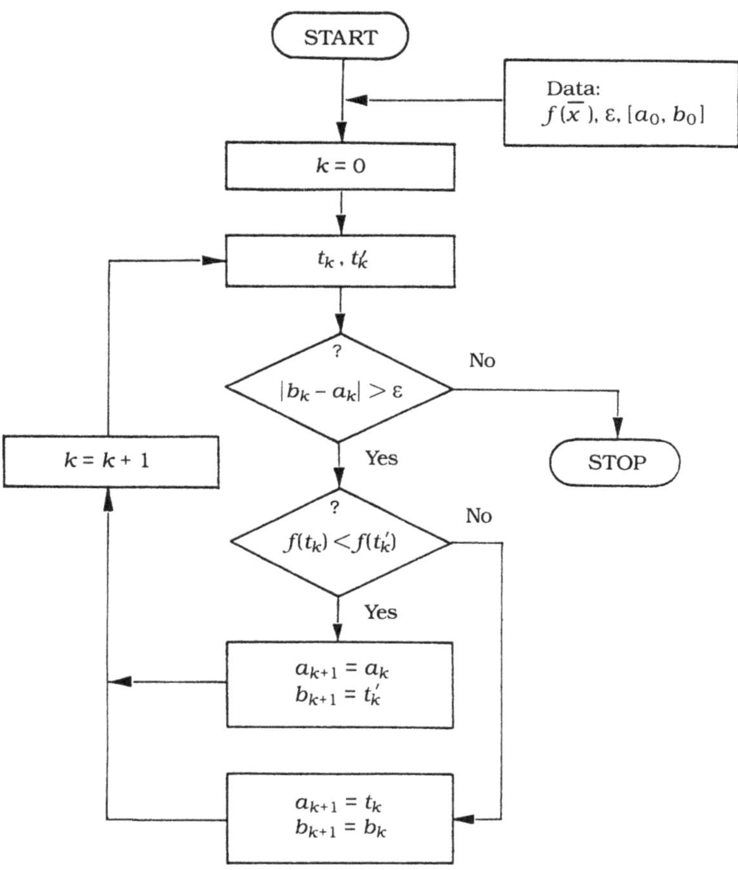

Figure 2.3 Flow diagram of the golden section method.

The definition of the gradient therefore points to the steepest ascent direction of function $f(\overline{x})$ in the neighborhood of point $\overline{x} = (x_1, x_2, \ldots, x_n)$.

Obviously, the vector $-\nabla f(\overline{x})$ is the direction of steepest descent, and is helpful in the search for the point at which the function $f(\overline{x})$ reaches its *minimum* value.

In most minimization problems, an analytic calculation of the partial derivatives is not possible in view of the complexity of the funciton $f(\overline{x})$. In such cases, an approximate numerical gradient can be calculated by using the following formulas:

$$\frac{\partial f}{\partial x_i} \approx f'_i = \frac{f(\overline{x} + \Delta x_i) - f(\overline{x})}{\Delta x_i} \tag{2.15}$$

$$\frac{\partial f}{\partial x_i} \approx f'_i = \frac{f(\bar{x} + \Delta x_i) - f(\bar{x} - \Delta x_i)}{2\,\Delta x_i} \tag{2.16}$$

The computation of the gradient by using (2.16) is more accurate than that using (2.15), but requires $2n/(n + 1)$ times more calculations of function $f(\bar{x})$.

The minimization methods for which the search direction from point $\bar{x}^k$ to point $\bar{x}^{k+1}$ is consistent with $-\nabla f(\bar{x}^k)$ are called *gradient methods.* These may have different search steps and stop criteria. In the simplest case, the move from $\bar{x}^k$ to $\bar{x}^{k+1}$ is in the direction $-\nabla f(\bar{x}^k)$ with a fixed step h, such that

$$\bar{x}^{k+1} = \bar{x}^k - h\frac{\nabla f(\bar{x}^k)}{|\nabla f(\bar{x}^k)|} \tag{2.17}$$

If $f(\bar{x}^{k+1}) > f(\bar{x}^k)$, $\bar{x}^{k+1} = \bar{x}^k$, and the search is performed in another direction, most often with the smaller step. A step proportional to the modulus of the gradient sometimes is more efficient, in effect,

$$\bar{x}^{k+1} = \bar{x}^k - h\nabla f(\bar{x}^k) \tag{2.18}$$

The typical stop criteria for the search of a minimum are

$$\begin{aligned} &|\nabla f(\bar{x})|\,\Big|_{\bar{x}=\bar{x}^s} < \epsilon \\ &f(\bar{x}^s) < f(\bar{x}^s + \Delta\bar{x}) \end{aligned} \tag{2.19}$$

where ϵ is a small positive value and $(\bar{x}^s + \Delta\bar{x})$ is a point in the neighborhood closest to the minimum $\bar{x}^s$.

2.3.1 The Steepest Descent Method

Let us consider the *steepest descent method* as the first gradient method, making use of the properties of the gradient direction and line search methods discussed in Section 2.2. After the gradient of function $f(\bar{x})$ at point $\bar{x}^k$ is calculated, the minimum is sought along the line [3].

$$\bar{x}^{k+1} = \bar{x}^k - t\nabla f(\bar{x}^k) \tag{2.20}$$

where t is a parameter. The minimum value of $f(\bar{x})$ in the direction determined by (2.20) can be found using the minimization techniques described in Section 2.2. Finding the minimum completes the search in this direction. The new search direction is determined by computing the gradient at the minimum, which has been found in the previous iteration. In this manner, the next (i.e., $(k + 1)$th) iteration begins.

The stop criterion in this method is

$$|\nabla f(\overline{x})| \leqslant \epsilon \tag{2.21}$$

where ϵ is a small positive value. In this case, we need to verify whether the solution found from (2.21) is the global minimum.

The flow diagram of the method under discussion is presented in Figure 2.4.

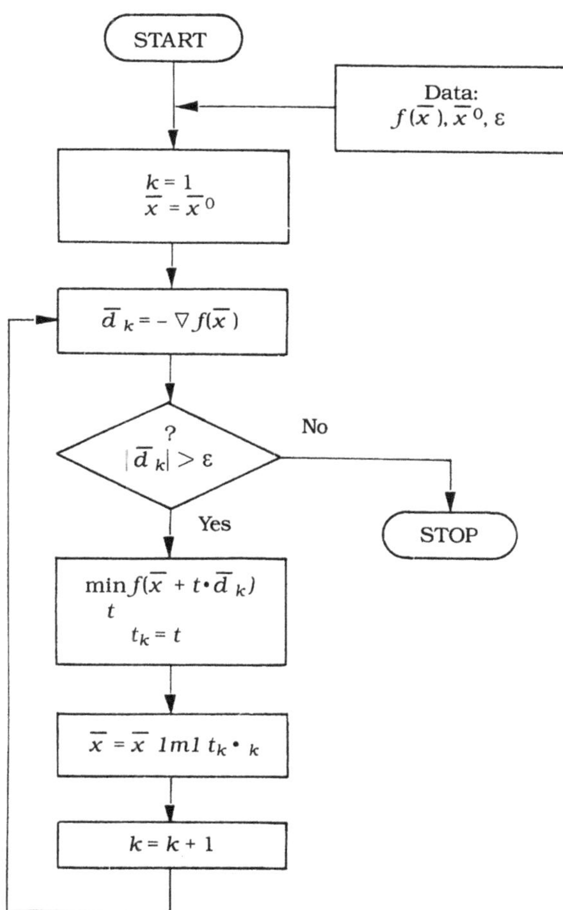

Figure 2.4 Flow diagram of the steepest descent method.

2.3.2 The Davidon-Fletcher-Powell Method

We know from numerical experiments that $-\nabla f(\bar{x})$ is the descent direction of function $f(\bar{x})$ only in the closest neighborhood of point $\bar{x}$ in which the function has been calculated. Therefore, more efficient minimization methods determine search directions with the help of the second derivatives or their approximations. The classical example of that group of methods is the Newton method, in which the search for the minimum is continued as follows:

$$\bar{x}^{k+1} = \bar{x}^k - H^{-1}(\bar{x}^k)\nabla f(\bar{x}^k) \tag{2.22}$$

where

$$H(\bar{x}^k) = \begin{bmatrix} \dfrac{\partial^2 f(\bar{x}^k)}{\partial^2 x_1}, \ldots, \dfrac{\partial^2 f(\bar{x}^k)}{\partial x_1 \partial x_n} \\ \cdots \qquad\qquad \cdots \\ \dfrac{\partial^2 f(\bar{x}^k)}{\partial x_n \partial x_1}, \ldots, \dfrac{\partial^2 f(\bar{x}^k)}{\partial^2 x_n} \end{bmatrix}$$

is the Hessian of function $f(\bar{x})$. Despite the quick convergence, this method is not widely used because we need the matrix of the second derivatives as well as its inverted form. Moreover, this method can be used only in the neighborhood closest to the minimum, where the Hessian matrix $H(\bar{x}^k)$ is positively determined.

The Newton method, however, cannot be neglected because it is the basis of many efficient minimization techniques described in the literature (see, e.g., [3,4]), such as the variable-metric methods. In these methods, the gradients computed in the previous iterations are used to calculate the approximation of the inverse of the Hessian matrix, denoted by $H^{-1}(\bar{x})$. For this purpose, a sequence of matrices D_k is created, which satisfies the condition:

$$D_k[\nabla f(\bar{x}^j) - \nabla f(\bar{x}^{j-1})] = \bar{x}^j - \bar{x}^{j-1} \tag{2.23}$$

for $j = 1, 2, \ldots, k$.

Comparison of (2.22) and (2.23) shows that, for the quadratic function $f(x_1, x_2, \ldots, x_n)$, after n steps in linearly independent directions, the matrices D_k and $H^{-1}(\bar{x}^k)$ are equivalent. The specific forms of calculating the matrices D_k may be different (see, for instance, the algorithms of Davidon-Fletcher-Powell, Wolfe-Broyden-Davidon, and Broyden-Goldfarb-Fletcher-Shanno methods described in [3]). For example, let us consider more precisely the Davidon-Fletcher-Powell method, which is widely used due to its efficiency and good convergence. Suppose that the initial point $\bar{x}^0$, from which the search for the minimum starts, is known. In the general case, the search direction is calculated as

$$\bar{d}^{k+1} = -D_k \nabla f(\bar{x}^k) \tag{2.24}$$

In the first iteration,

$$\bar{d}^1 = -D_0 \nabla f(\bar{x}^0)$$

where D_0 denotes the unit matrix. Point $\bar{x}^1$ in which function $f(\bar{x})$ reaches its minimum is sought along the direction $\bar{d}^1$ by one of the one-dimensional minimization techniques described in Section 2.2 (for instance, the golden section method). In the next iterations, matrix D_{k+1} is calculated as follows [6]:

$$D_{k+1} = D_k + A_k - B_k \tag{2.25}$$

where

$$A_k = \frac{\bar{s}^{k+1}[\bar{s}^{k+1}]^T}{[\bar{s}^{k+1}]^T \bar{r}^{k+1}}$$

$$B_k = \frac{D_k \bar{r}^{k+1}[\bar{r}^{k+1}]^T D_k^T}{[\bar{r}^{k+1}]^T D_k \bar{r}^{k+1}}$$

$$\bar{s}^{k+1} = \bar{x}^{k+1} - \bar{x}^k$$

$$\bar{r}^{k+1} = \nabla f(\bar{x}^{k+1}) - \nabla f(\bar{x}^k)$$

When matrix D_{k+1} is known, the next search direction, namely $\bar{d}^{k+2}$, and the minimum point $\bar{x}^{k+2}$ corresponding to this direction, are found in the same way.

A significant feature of this method is its ability to start again, in which case matrix D_k is substituted by the unit matrix. Consequently the revised search direction becomes

$$\bar{d}^{k+1} = -\nabla f(\bar{x}^k)$$

The ability to start again is necessary when the search along a final direction does not ensure finding the minimum or when $2n + 1$ iterations have been made.

The search for the minimum terminates if (2.19) is satisfied.

The flow diagram of this method is shown in Figure 2.5. Moreover, this method is illustrated by the example given below.

Example

Let us consider the following function of two real variables:

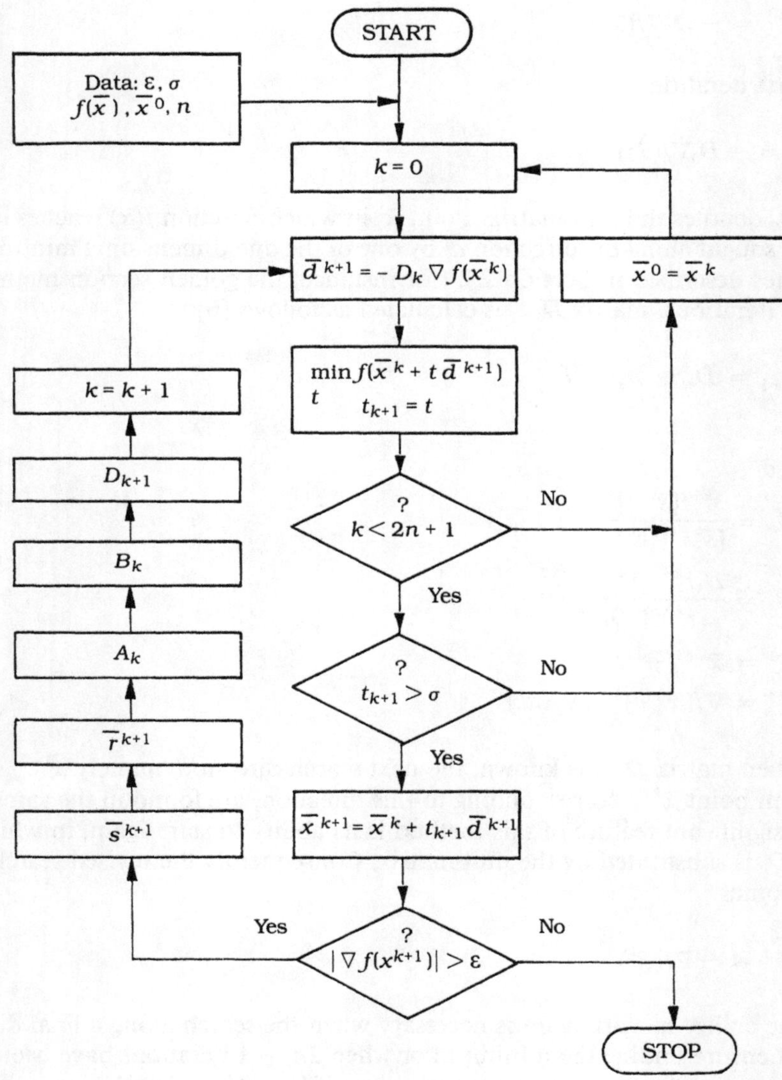

Figure 2.5 Flow diagram of the Davidon-Fletcher-Powell minimization strategy.

$$f(\overline{x}) = f(x_1, x_2) = 3(x_1 - 2)^2 + (x_2 - 1)^2$$

which has the global minimum at point (2, 1). Let $\overline{x}^0 = [3, 3]$ be the starting point, where $f(\overline{x}^0) = 7$.

First Iteration

The direction $\overline{d}^1$ calculated at point (3, 3) is equal to

$$\nabla f(\overline{x}^0) = [6(x_1 - 2), 2(x_2 - 1)]^T \Big|_{\substack{x_1=3 \\ x_2=3}} = [6, 4]^T$$

Next, the minimum is sought from the formula:

$$\overline{x}^1 = \overline{x}^0 - t_1 D_0 \nabla f(\overline{x}^0)$$

where the components of $\overline{x}^1$ are functions of the parameter t_1. Thus,

$$\begin{bmatrix} x_1^1 \\ x_2^1 \end{bmatrix} = \begin{bmatrix} 3 \\ 3 \end{bmatrix} - t_1 \begin{bmatrix} 1, 0 \\ 0, 1 \end{bmatrix} \begin{bmatrix} 6 \\ 4 \end{bmatrix} = \begin{bmatrix} 3 - 6t_1 \\ 3 - 4t_1 \end{bmatrix}$$

After substitution in $f(x_1, x_2)$ of x_1 and x_2 by $x_1^1(t_1)$ and $x_2^1(t_1)$, we obtain

$$f(t_1) = 3(1 - 6t_1)^2 + (2 - 4t_1)^2$$

The value of parameter t_1 for which the function $f(t_1)$ reaches its minimum can be calculated from the condition:

$$\frac{df(t_1)}{dt_1} = 0$$

In this case, function $f(t_1)$ takes its minimum for $t_1 = 0.209677$. Finally, $\overline{x}^1 = [1.741935, 2.161290]$, $f(\overline{x}^1) = 1.548387$, and $\nabla f(\overline{x}^1) = [-1.548390, 2.322580]^T$.

Second Iteration

To obtain the second search direction $\overline{d}^2$, we calculate the vectors, $\overline{r}^1$ and $\overline{s}^1$, as

$$\overline{r}^1 = \nabla f(\overline{x}^1) - \nabla f(\overline{x}^0) = [-7.548390, -1.677420]^T$$
$$\overline{s}^1 = \overline{x}^1 - \overline{x}^0 = [-1.258065, -0.838710]^T$$

These vectors are used to calculate the matrix:

$$D_1 = D_0 + A_0 - B_0$$

where

$$D_0 = \begin{bmatrix} 1, & 0 \\ 0, & 1 \end{bmatrix}$$

$$A_0 = \frac{\begin{bmatrix} -1.258065, & 0 \\ -0.838710, & 0 \end{bmatrix} \begin{bmatrix} -1.258065, & -0.838710 \\ 0, & 0 \end{bmatrix}}{[-1.258065, -0.838710][-7.548390, -1.677420]^T}$$

$$= \begin{bmatrix} 0.145161, 0.096774 \\ 0.096774, 0.064516 \end{bmatrix}$$

$$B_0 = \frac{\begin{bmatrix} 1, & 0 \\ 0, & 1 \end{bmatrix} \begin{bmatrix} -7.548390, & 0 \\ -1.677420, & 0 \end{bmatrix} \begin{bmatrix} -7.548390, & -1.677420 \\ 0, & 0 \end{bmatrix} \begin{bmatrix} 1, & 0 \\ 0, & 1 \end{bmatrix}}{[-7.548390, -1.677420] \begin{bmatrix} 1, & 0 \\ 0, & 1 \end{bmatrix} \begin{bmatrix} -7.548390 \\ -1.677420 \end{bmatrix}}$$

$$= \begin{bmatrix} 0.952941, & 0.211764 \\ 0.211764, & 0.047058 \end{bmatrix}$$

Here, the superscript T denotes the transpose of a matrix, e.g., $[-7.548390, -1.677420]^T = \begin{bmatrix} -7.548390 \\ -1.677420 \end{bmatrix}$. In the above relations, the premultiplications of a column vector by a row vector of the same dimension are performed according to the following standard rule:

$$\begin{bmatrix} a_1 \\ a_2 \\ \vdots \\ a_n \end{bmatrix} [b_1, b_2, \ldots, b_n] = \begin{bmatrix} a_1b_1, a_1b_2, \ldots, a_1b_n \\ a_2b_1, a_2b_2, \ldots, a_2b_n \\ \ldots \\ a_nb_1, a_nb_2, \ldots, a_nb_n \end{bmatrix}$$

The particular elements of matrix D_1 are computed by summation of the corresponding elements of matrices D_0, A_0 and $-B_0$. Thus,

$$D_1 = \begin{bmatrix} 0.192220, & -0.114990 \\ -0.114990, & 1.017458 \end{bmatrix}$$

The components of $\bar{x}^2$ are calculated similarly as in the first iteration, namely,

$$\bar{x}^2 = \bar{x}^1 - t_2 D_1 \nabla f(\bar{x}^1)$$

The extended form of this relation is

$$\begin{bmatrix} x_1^2 \\ x_2^2 \end{bmatrix} = \begin{bmatrix} 1.741935 \\ 2.161290 \end{bmatrix} - t_2 \begin{bmatrix} 0.192220, & -0.114990 \\ -0.114990, & 1.017458 \end{bmatrix} \begin{bmatrix} -1.548390 \\ 2.322580 \end{bmatrix}$$

After multiplication, we have $x_1^2 = 1.741935 + 0.564705\ t_2$ and $x_2^2 = 2.161290 - 2.541176\ t_2$. By substitution of x_1 and x_2 in $f(x_1, x_2)$ by $x_1^2(t_2)$ and $x_2^2(t_2)$, we obtain function $f(t_2)$, which reaches its minimum for $t_2 = 0.456988$. Point $\overline{x}^2$ at which function $f(\overline{x})$ reaches its minimum in the direction $\overline{d}^2$ is [1.999998, 1.000002], which means that the global minimum has been found.

2.3.3 The Fletcher-Reeves Method

An important group of gradient minimization techniques is the *conjugate gradient* approach [3,7]. Therefore, one of these techniques, the Fletcher-Reeves method, is presented below. The flow diagram of this method is shown in Figure 2.6. In the general case, the algorithm consists of the following steps:

1. Determination of the initial point $\overline{x}^0$ and the small positive number ϵ, which is a stop criterion parameter.
2. Setting $v = 0$.
3. Computation of $\nabla f(\overline{x}^v)$.
4. Determination of the search direction $\overline{d}^{v+1} = -\nabla f(\overline{x}^v)$.
5. Finding the value of parameter t for which function $f(\overline{x}^v + t\overline{d}^{v+1})$ reaches its minimum.
6. Checking to see if the number of iterations is equal to n, where n is the dimension of vector $\overline{x}$. If so, the search must be restored by assuming $\overline{x}^0 = \overline{x}^v$ and returning to step 2.
7. Stopping if $|\nabla f(\overline{x}^{v+1})| \leqslant \epsilon$, otherwise proceeding to the next step.
8. Calculation of the auxiliary coefficient:

$$p_{v+1} = \frac{[\nabla f(\overline{x}^{v+1})]^T[\nabla f(\overline{x}^{v+1}) - \nabla f(\overline{x}^v)]}{[\nabla f(\overline{x}^v)]^T \nabla f(\overline{x}^v)}$$

9. Calculation of the new search direction:

$$\overline{d}^{v+2} = -\nabla f(\overline{x}^{v+1}) + p_{v+1}\overline{d}^{v+1}$$

10. Setting $v = v + 1$ and returning to step 5.

The algorithm presented above is illustrated by the following example.

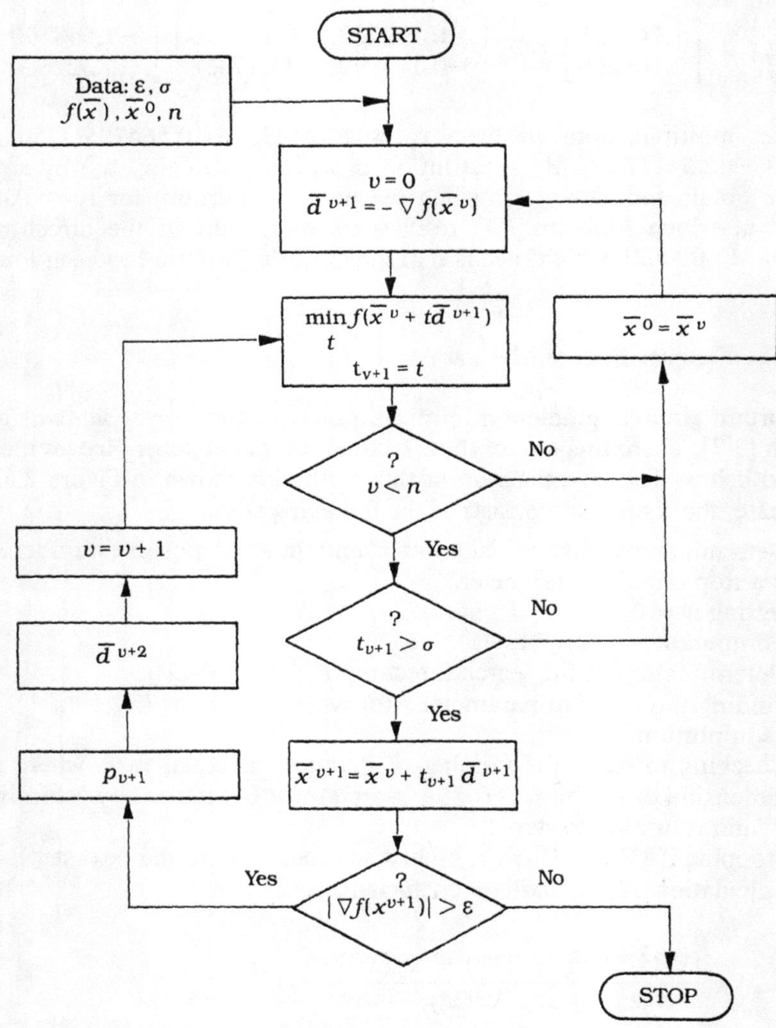

Figure 2.6 Flow diagram of the Fletcher-Reeves minimization method.

Example

Let us consider again the objective function:

$$f(\bar{x}) = f(x_1, x_2) = 3(x_1 - 2)^2 + (x_2 - 1)^2$$

which reaches the global minimum at point $\overline{x}^s = [2, 1]$.
Let $\overline{x}^0 = [3, 3]$ be used as the initial point, where $f(\overline{x}^0) = 7$.

First Iteration

The search direction $\overline{d}^1$ is given by

$$-\nabla f(\overline{x}^0) = [-6(x_1 - 2), -2(x_2 - 1)]^T \Big|_{\substack{x_1=3 \\ x_2=3}} = [-6, -4]^T$$

Hence,

$$\begin{bmatrix} x_1^1 \\ x_2^1 \end{bmatrix} = \begin{bmatrix} 3 \\ 3 \end{bmatrix} - t_1 \begin{bmatrix} 6 \\ 4 \end{bmatrix} = \begin{bmatrix} 3 - 6t_1 \\ 3 - 4t_1 \end{bmatrix}$$

After substituting for x_1 and x_2 in $f(x_1, x_2)$ by $x_1^1(t_1)$ and $x_2^1(t_1)$, we obtain:

$$f(t_1) = 3(1 - 6t_1)^2 + (2 - 4t_1)^2$$

The value of parameter t_1, for which function $f(t_1)$ reaches its minimum, can be found from the condition:

$$\frac{df(t_1)}{dt_1} = 0$$

This condition is satisfied for $t_1 = 0.209677$, and consequently $\overline{x}^1 = [1.741935, 2.161290]$, $\nabla f(\overline{x}^1) = [-1.548390, 2.322580]^T$, and $p_1 = 0.149844$.

Second Iteration

The search direction $\overline{d}^2$ is calculated with the help of $\nabla f(\overline{x}^1)$, $\overline{d}^1$, and p_1 obtained in the previous iteration:

$$\overline{d}^2 = -\nabla f(\overline{x}^1) + p_1\overline{d}^1 = [0.649323, -2.921956]^T$$

As in the first iteration,

$$\begin{bmatrix} x_1^2 \\ x_2^2 \end{bmatrix} = \begin{bmatrix} 1.741935 \\ 2.161290 \end{bmatrix} + t_2 \begin{bmatrix} 0.649323 \\ -2.921956 \end{bmatrix}$$

$$= \begin{bmatrix} 1.741935 + 0.649323t_2 \\ 2.161290 - 2.921956t_2 \end{bmatrix}$$

After substituting for x_1 and x_2 in function $f(x_1, x_2)$ by $x_1^2(t_2)$ and $x_2^2(t_2)$, we obtain

$$f(t_2) = 3(-0.258065 + 0.649323t_2)^2 + (1.161290 - 2.921956t_2)^2$$

The function $f(t_2)$ takes its minimum for $t_2 = 0.397436$. In this case, $\overline{x}^2 = [2, 1]$, $f(\overline{x}^2) = 0$, $\nabla f(\overline{x}^2) = [0, 0]^T$, and $p_2 = 0$. This completes the search for the global minimum.

2.4 PENALTY FUNCTION METHODS

The methods described in Section 2.3 are examples of classical minimization techniques for problems without constraints. These methods also may be used, after some modification, for solving minimization problems with linear and nonlinear constraints. Therefore, in this section, we discuss a transformation of the constrained minimization problem into an equivalent sequence of unconstrained minimization problems. The penalty function methods based on this transformation generally have simple mathematical forms, and for this reason can be utilized in most practical cases without difficulty.

Hence, let us consider the minimization of the following objective function:

$$f(\overline{x}) = (x_1 - 3)^2 + (x_2 - 2)^2 \tag{2.26}$$

subject to the constraint:

$$h(\overline{x}) = x_1 + x_2 = 0$$

The constrained minimization problem may be transformed into an equivalent unconstrained minimization problem by adding a suitably constructed penalty to the original objective function, for instance,

$$p(\overline{x}) = f(\overline{x}) + h^2(\overline{x}) \tag{2.27}$$

In this case,

$$p(\overline{x}) = 2x_1^2 + 2x_2^2 - 6x_1 - 4x_2 + 2x_1x_2 + 13$$

During the minimization of function $p(\overline{x})$, the penalty term $h^2(\overline{x})$ causes the solution to reach the feasible region, where $h(\overline{x}) = 0$. Usually, the purpose of penalty functions is to bring the solution into the feasible region determined by the given constraints. In the penalty function approach, the constraints are not explicitly considered; therefore, the minimum of the modified extended objective function may be sought by using the simpler and more efficient unconstrained minimization methods, such as described in Section 2.3.

In the literature (see, e.g., [3,4]) penalty function methods are frequently classified into two groups, exterior and interior. The essence of these minimization techniques is presented below.

2.4.1 Exterior Penalty Function Methods

Let us consider the minimization of the objective function $f(\overline{x})$ in the feasible region R^n determined by the inequalities:

$$r_i(\overline{x}) \leq 0, \text{ for } i = 1, 2, \ldots, n \tag{2.28}$$

The modified objective function $p(\overline{x})$ is obtained by adding a penalty term $q(\overline{x})$, which satisfies the following conditions:

$q(\overline{x}) = 0$, inside the feasible region R^n

$q(\overline{x}) > 0$, if at least one of inequalities is not satisfied

The following expression may be an example of such a penalty function:

$$q(\overline{x}) = \sum_{i=1}^{n} \left[\frac{r_i(\overline{x}) + |r_i(\overline{x})|}{2} \right]^2 \tag{2.29}$$

Hence, for a given penalty function $q(\overline{x})$, the modified objective function is

$$p(\overline{x}) = f(\overline{x}) + tq(\overline{x}) \tag{2.30}$$

where t is the auxiliary positive parameter.

The minimization technique is based on a sequence of penalty functions given by $t_k q(\overline{x})$, where t_k is the discrete set of monotonically increasing values. For the initial element, t_1, of this set, the minimum of function $p(\overline{x}, t_1)$ (see (2.30)) can be evaluated by means of the methods described in Section 2.3. Let $\overline{x}^1$ be the solution of this minimization problem, which is taken as an initial point for minimization of the next function $p(\overline{x}, t_2)$, where $t_2 > t_1$. As a result, we solve the sequence of minimization problems with functions $p(\overline{x}, t_k)$, where $k = 1, 2, \ldots$. If the value of parameter t_k increases, the minima of functions $p(\overline{x}, t_k)$ converge to the minimum of the original objective function $f(\overline{x})$, which, of course, lies in the feasible region R^n.

Now let us consider the problem with additional constraints given in the form of equalities:

$$h_m(\overline{x}) = 0, \text{ for } m = n + 1, n + 2, \ldots, p$$

For this case, the corresponding component of the penalty function can be assumed as

$$e(\overline{x}) = \sum_{m=n+1}^{p} |h_m(\overline{x})|^{\alpha} \tag{2.31}$$

where α is a positive parameter. Finally, the modified objective function becomes

$$p(\overline{x}, t) = f(\overline{x}) + tq(\overline{x}) + e(\overline{x}) \tag{2.32}$$

The coefficient α should be chosen in such a way that the penalty does not increase rapidly outside the feasible region, particularly during the initial stage of the search. From a practical point of view the initial point $\overline{x}^0$ must be either inside or outside the feasible region. The efficiency of this method significantly depends on the choice of the initial point $\overline{x}^0$ as well as on the values of coefficients t and α. In order to illustrate the present method, let us consider the following example in which the initial point is outside the feasible region.

Example

Let us consider the minimization of the objective function:

$$f(\overline{x}) = f(x_1, x_2) = 3(x_1 - 2)^2 + (x_2 - 1)^2$$

subject to the constraints:

$$x_1 \geqslant 0.5$$
$$x_2 \geqslant 0$$
$$x_1 + x_2 - 2 \leqslant 0$$

This problem can be reformulated in the following standard form:

$$\min_{\overline{x} \subset R^2} f(\overline{x})$$

$$R^2 = \begin{cases} x_1 + x_2 - 2 \leq 0 \\ -x_1 + 0 + 0.5 \leq 0 \\ 0 - x_2 + 0 \leq 0 \end{cases}$$

After assuming that the penalty term $q(\overline{x})$ has the form (2.29), we obtain the following modified objective function:

$$p(\overline{x}, t) = 3(x_1 - 2)^2 + (x_2 - 1)^2 + t\left[\left(\frac{-x_2 + |-x_2|}{2}\right)^2 + \left(\frac{x_1 + x_2 - 2 + |x_1 + x_2 - 2|}{2}\right)^2 + \left(\frac{-x_1 + 0.5 + |-x_1 + 0.5|}{2}\right)^2\right]$$

The function $p(\overline{x}, t)$ is equivalent to $f(\overline{x})$ for $\overline{x} \subset R^2$, where R^2 is the feasible region (see Figure 2.7). Let us assume that the above minimization problem is being

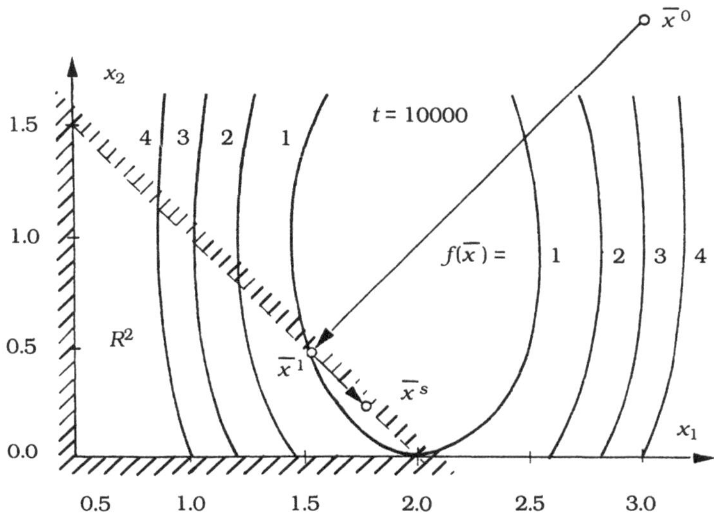

Figure 2.7 Optimization problem with three inequality constraints.

solved by the Fletcher-Reeves method. In this case, the following derivatives are helpful:

$$\frac{dp(\overline{x}, t)}{dx_1} = 6(x_1 - 2) + t[(x_1 + x_2 - 2 + |x_1 + x_2 - 2|)$$

$$-(-x_1 + 0.5 + |-x_1 + 0.5|)]$$

$$\frac{dp(\overline{x}, t)}{dx_2} = 2(x_2 - 1) + t[(x_1 + x_2 - 2 + |x_1 + x_2 - 2|)$$

$$-(-x_2 + |-x_2|)]$$

If

$$\overline{d}^j = [dx_1^j, dx_2^j]^T$$

is the search direction in the jth iteration, the components x_1^j and x_2^j of the minimum $\overline{x}^j$ are

$$\begin{bmatrix} x_1^j \\ x_2^j \end{bmatrix} = \begin{bmatrix} x_1^{j-1} \\ x_2^{j-1} \end{bmatrix} + h_{min}^j \begin{bmatrix} dx_1^j \\ dx_2^j \end{bmatrix}$$

The results of minimization of function $p(\overline{x}, t)$ obtained for the different values of the penalty coefficient t and the same starting point $\overline{x}^0 = [3, 2]$ are given in Tables 2.1, 2.2, and 2.3.

Table 2.1

$t = 0$

j	dx_1^j	dx_2^j	h_{min}^j	x_1	x_2	$p(\overline{x}, t)$
1	−6	−2	0.178571	1.928	1.643	0.428
2	0.153061	−1.377551	0.466666	2.000	1.000	0.000
3	$-5.4\ 10^{-9}$	$-1.8\ 10^{-9}$	0.172727	2.000	1.000	0.000

Table 2.2

$t = 100$

j	dx_1^j	dx_2^j	h_{min}^j	x_1	x_2	$p(\overline{x}, t)$
1	−606	−602	0.002475	1.500	0.509	0.999
2	1.004797	−1.014861	0.250625	1.751	0.255	0.744
3	$-1.3\ 10^{-7}$	$-1.2\ 10^{-7}$	0.002462	1.751	0.255	0.744

Table 2.3

$t = 10000$						
j	dx_1^j	dx_2^j	h_{min}^j	x_1	x_2	$p(\overline{x}, t)$
1	−600006	−60002	0.000025	1.500	0.500	1.000
2	1.000050	−1.000151	0.250006	1.750	0.250	0.749
3	−44.7 10^{-6}	−44.7 10^{-6}	25.1 10^{-6}	1.750	0.250	0.750

2.4.2 Interior Penalty Function Methods

In these methods, the modified objective functions include the components called *interior penalty functions* or *barriers.* The values of these components increase to infinity when the solution approaches the boundary of the feasible region. Therefore, the interior penalty function methods are preferred for solving problems where the constraints are given in the form of inequalities:

$$r_i(\overline{x}) \leqslant 0, \text{ for } i = 1, 2, \ldots, n$$

The classical examples of the interior penalty functions (barriers) are

$$i_1(\overline{x}) = -\sum_{i=1}^{n} \ln\,[-r_i(\overline{x})] \tag{2.33}$$

and

$$i_2(\overline{x}) = -\sum_{i=1}^{n} 1/r_i(\overline{x})$$

When the form of an interior penalty function is known, the modified objective function is

$$p(\overline{x}, t) = f(\overline{x}) + ti(\overline{x}) \tag{2.34}$$

where $f(\overline{x})$ is the original objective function, and t is the auxiliary positive parameter.

In the minimization process, a sequence of monotonically decreasing positive values t_k, where $k = 1, 2, \ldots,$ is used. Any interior feasible point can be used as the initial (starting) point. The minimum $\overline{x}^1$ of function $p(\overline{x}, t_1)$ is the starting point for the next minimization problem, $p(\overline{x}, t_2)$, and so on. The sequence of the minimal points $\overline{x}^k$ is convergent to the solution of the original minimization problem ($t_k \approx 0$).

In general, this group of methods ensures that the search for the minimum is done inside the feasible region. If the value of parameter t decreases, the influence of the penalty component on $p(\overline{x}, t)$ also decreases, and the influence of the true (original) objective function $f(\overline{x})$ increases. Therefore, the sequence of minimization problems $p(\overline{x}, t_k)$ leads to a sufficiently good approximation of the minimum of the original problem given by objective function $f(\overline{x})$ and the constraints related to it.

REFERENCES

[1] Fletcher, R., *A Review of Methods for Unconstrained Optimization,* New York: Academic Press, 1969.

[2] Zangwill, W.I., *Nonlinear Programming: A Unified Approach,* Englewood Cliffs, NJ: Prentice-Hall, 1969.

[3] Himmelblau, D.M., *Applied Nonlinear Programming,* New York: McGraw-Hill, 1972.

[4] Bazaraa, M.S., and C.M. Shetty, *Nonlinear Programming: Theory and Algorithms,* New York: John Wiley and Sons, 1979.

[5] Bandler, J.W., *"Optimization Methods for Computer-Aided Design," IEEE Trans. Microwave Theory and Techniques,* Vol. MTT-17, No. 8, August 1969, pp. 533–552.

[6] Fletcher, R. and M.J.D. Powell, "A Rapidly Convergent Descent Method for Minimization," *The Computer Journal,* Vol. 6, No. 2, February 1963, pp. 163–168.

[7] Fletcher, R., and C.M. Reeves, "Function Minimization by Conjugate Gradients," *The Computer Journal,* Vol. 7, No. 7, July, 1964, pp. 149–154.

[8] Fiacco, A.V., and G.P. McCormick, *Nonlinear Programming: Sequential Unconstrained Minimization Techniques,* New York: John Wiley and Sons, 1968.

Chapter 3

CHARACTERISTICS OF THE "MICRO" DESIGN SYSTEM

3.1 GENERAL DESCRIPTION

Computer-aided design in its strict interpretation may be treated as any design process in which a computer is used as a tool. With regard to the microwave circuits, this process generally consists of the following important stages:

1. Synthesis of an electrical circuit with lumped or distributed parameters which satisfies all the required electrical specifications;
2. Optimization of the electrical circuit with respect to its important parameters, for instance optimization of its insertion loss function;
3. Realization of the circuit being designed by evaluating its main structural parameters, such as the general layout or geometrical dimensions;
4. Verification of the structural constraints resulting from the assumed realization technique. If these constraints are not satisfied, the electrical scheme is modified in a suitable manner, and after this the remaining design stages are performed again.

In each of the stages, we should try to use the computer rationally so as to make the whole design process more accurate, reliable, and fast, and to eliminate human effort as much as possible [5,7].

The design process is rather complicated and, for its effective realization we apply specialized programs running on powerful computers. Unfortunately, many engineers have limited access to such advanced design systems. Further, to apply advanced general-purpose design systems is not always necessary. In many cases, the approach and design routines presented here will be sufficient.

As we know from the literature (see, e.g., [1–9]), many linear microwave circuits, even the most complicated, can be treated as combinations of basic components, such as attenuators, impedance transformers, directional couplers, and different types of filters. These basic components, sometimes called *building blocks,* can be characterized in terms of the voltages and currents at their ports, as shown in Figure 3.1. At microwave frequencies, the voltages, currents, and impedances are often replaced by normalized wave variables described in Sections 4.3 and 4.4, and the resulting scattering matrix formulation is then used [5,7,8].

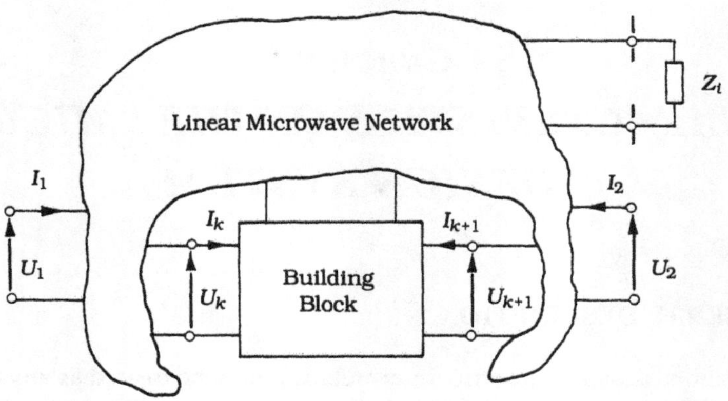

Figure 3.1 The multiport linear microwave network composed of many building blocks.

Let us assume that the overall electrical structure of the network being designed is known, and that the electrical specifications for the particular basic components have been determined. Under this assumption, the individual basic components, isolated from the whole network, can be designed separately to meet their respective electrical specifications.

As a rule, these component circuits have simpler electrical structures, and therefore they can be designed by using less complicated computer routines. Also, in these cases, the overall design process consists of the stages mentioned earlier in points 1, 2, 3, and 4. This means that in the design process we must use different kinds of computer programs.

According to the design scheme, these programs should allow us to synthesize and optimize the electrical circuit, and to evaluate the corresponding structural parameters of the device being designed.

To design in this way is possible by using the MICRO system, the organizational diagram of which is shown in Figure 3.2. It consists of three groups of computational programs included in Libraries L1, L2, and L3. All these programs are specified in the initial procedure, MENU, and can be selected directly from it or indirectly by means of the control procedure, CONTROL (see Figure 3.2).

The programs included in Library L1 create a mathematical basis for the system. The programs are used to transform the electrical parameters of the circuit

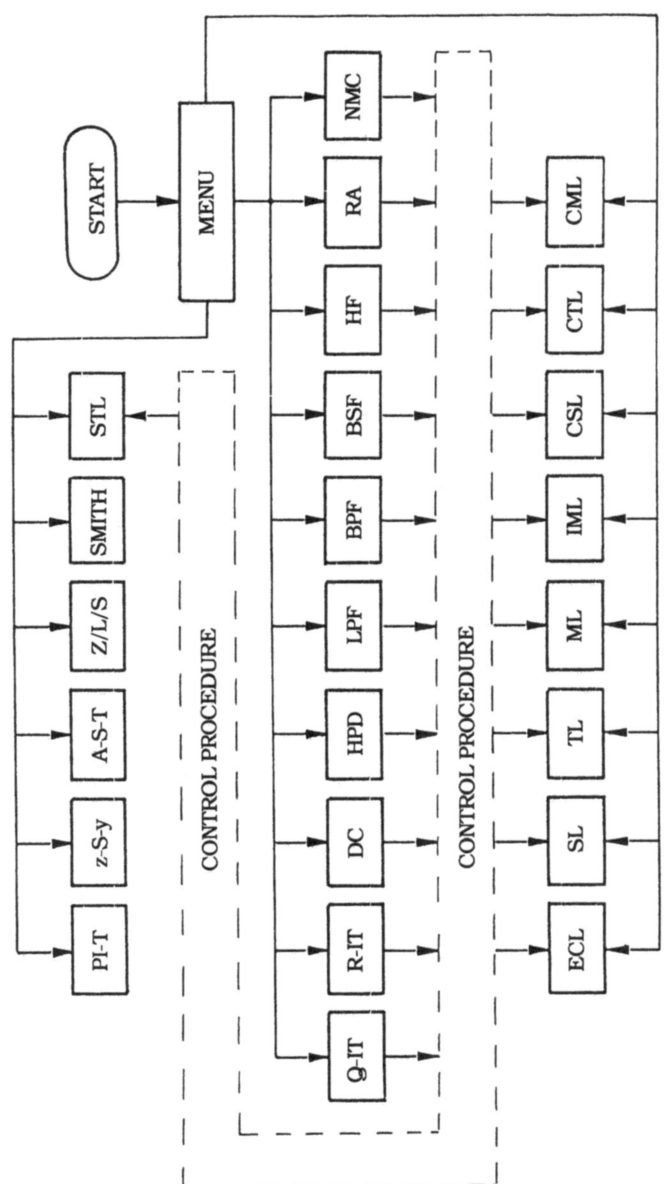

Figure 3.2 The organization of the MICRO design system.

under design to a form convenient for the next stages of the design process (i.e., they allow transformations between different circuit representations, as described in Chapter 4). The name and purpose of each program is given in Table 3.1. The mathematical essentials of the problems presented in Table 3.1 are given together with the corresponding BASIC listings and check calculations in Chapter 4.

Table 3.1
Computer Programs of Library L1

Name	*Purpose*
PI-T	Transformation from section π to T and *vice versa*
z-S-y	Reciprocal transformations of matrices z, S and y of a two-port electrical network
A-S-T	Reciprocal transformations of matrices *ABCD*, S and T of a two-port network
Z/L/S	Some useful S-parameter relationships relevant to the analysis of two-port electrical networks
SMITH	Transformation of a terminating complex impedance via the lossy multisection stepped transmission line
STL	Design of the stepped transmission line on the basis of the insertion loss function (described in detail in Section 5.3)

The programs of the second group, included in Library L2, are intended for designing standard passive linear microwave circuits and their basic components. The name and purpose of each program is given in Table 3.2. Their mathematical bases, flow charts, BASIC listings, and check calculations are presented in Chapter 5.

Table 3.2
Computer Programs of Library L2

Name	*Purpose*
Q-IT	Design of quarter-wavelength stepped impedance transformers
R-IT	Design of monotonic and nonmonotonic stepped impedance transformers
DC	Design of microwave directional couplers
HPD	Design of three-port hybrid power dividers
LPF	Design of low-pass filters with lumped and distributed parameters
BPF	Design of bandpass filters with distributed parameters
BSF	Design of bandstop filters with distributed parameters
HF	Design of the multiharmonic filter
RA	Design of resistance matching attenuators
NMC	Design of narrow-band impedance matching circuits with lumped and distributed parameters

Library L3, including programs for designing some practically important single and coupled transmission lines, is organized in a similar manner. Brief characteristics and more detailed descriptions of these programs are presented in Table 3.3 and Chapter 6, respectively. The links between the calling (see Library L2) and called programs are compiled in Table 3.4. The programs included in Libraries L1 and L3 can also be called directly by commands given in the MENU.

Table 3.3
Computer Programs of Library L3

Name	*Purpose*
ECL	Design of the eccentric coaxial line
SL	Design of the slab line
TL	Design of the triplate stripline
ML	Design of the microstrip line
IML	Design of the inverted microstrip line
CSL	Design of coupled slab lines
CTL	Design of edge coupled triplate striplines
CML	Design of coupled microstrip lines

Table 3.4
Permissible Connections between Computer Programs of the MICRO Design System

Programs	*ECL*	*SL*	*TL*	*ML*	*IML*	*CSL*	*CTL*	*CML*
Q-IT	+	+	+	+	+			
R-IT	+	+	+	+	+			
DC		+	+	+		+	+	+
HPD		+		+				
LPF	+		+	+				
BPF			+	+			+	+
BSF		+	+	+				
HF	+		+					
RA								
NMC	+	+	+	+				

In creating the system, care has been taken to ensure that it is open, universal, and simple (in terms of its organization) so that it can be understood by any user. In most of the programs presented, in addition to computational statements, there are also instructions for avoiding different kinds of errors. For instance, calculations will not be made for incorrect input data, and the appropriate report is printed. Similarly, errors due to mathematical operations (for instance, dividing by zero), although they are not very probable, are eliminated by means of suitable check instructions.

The openness of the system allows for the possibility of expansion by adding new programs to the individual libraries. Moreover, the computer design system should be as universal as possible. In fact, it defines a more general system with programs that make possible the design of different electrical circuits satisfying similar technical requirements. For instance, program DC serves to design the $3\lambda/2$ hybrid-ring, two-branch, one-section, quarter-wavelength, and asymmetric high-pass TEM-mode directional couplers. An adequate structure for the directional coupler can be chosen manually by the user or automatically by the computer on the basis of input data and decisional criteria implemented in the program.

The following examples explain the essence of protection procedures included in the programs. Let us suppose that the hybrid-ring directional coupler (see Section 5.4) is being designed. In this case, the program CSL cannot be called as a subroutine because directional couplers of this type are realized only by segments of single (uncoupled) transmission lines. When an invalid command or variable is introduced into the computer memory as input data, the computer reports that execution of the program is impossible, and presents the reason for this decision. Other protection procedures are related with the conditions of physical realization. The protection is included in some programs to avoid useless calculations. For example, when designing the matching attenuator (see Section 5.10), the calculations will not be performed if the given attenuation L is less than the minimum attenuation L_{min}, due to the matching condition. Otherwise, some resistances of the attenuator being considered would be negative, which, of course, excludes such a solution as useless in practice. In this and similar cases, the computer informs us why the calculations have not been made.

To make the design system more clear, all of its computational programs have been written in BASIC, and simplifications and software tricks that can make the computations less straightforward have been avoided.

3.2 THE CONTROL PROGRAM

The control unit of the design system is a program consisting of two procedures, MENU and CONTROL (see Figure 3.2). The procedure MENU presents some possibilities for the whole system, and informs us how to use the particular programs included in Libraries L1, L2, and L3. Table 3.5 shows how these programs are distributed in the computer memory.

The instructions for CONTROL are located mainly in programs of Library L2. The instructions determine which programs from Libraries L1 and L3 can be called subroutines. In addition, this procedure also contains some decisional criteria due to structural limitations. In general, these criteria eliminate solutions that are practically difficult to realize.

Table 3.5
Addresses of the MICRO Computer Programs and Their Subroutines

Programs		*Memory Space*	*Subroutines*
Library L1	PI-T	200 to 328	
	z-S-y	400 to 562	
	A-S-T	700 to 966	
	Z/L/S	1100 to 1322	
	SMITH	1400 to 1552	
	STL	1600 to 2374	1688
Library L2	Q-IT	2500 to 2896	
	R-IT	3000 to 3382	
	DC	3500 to 4240	
	HPD	4400 to 4724	
	LPF	4800 to 5188	
	BPF	5300 to 5728	
	BSF	6200 to 6466	
	HF	6600 to 6742	
	RA	6800 to 6914	
	NMC	7000 to 7532	
Library L3	ECL	7700 to 7800	7742
	SL	7900 to 8038	7972
	TL	8200 to 8300	8252
	ML	8400 to 8588	8454
	IML	8700 to 8818	8750
	CSL	8900 to 9076	8966
	CTL	9300 to 9486	9360
	CML	9600 to 9876	9670

Listing of Program MENU

```
10 REM MENU
12 CLS:CLEAR
14 PRINT "MICRO - microwave design kit, version 2.0"
16 PRINT
18 PRINT "by  Stanislaw Rosloniec
20 PRINT "Department of Electronic Engineering"
22 PRINT "Warsaw Technical University"
24 PRINT "Nowowiejska 15/19"
26 PRINT "00-665 Warsaw, POLAND"
28 PRINT
30 PRINT "The presented here design kit is an integral part of
the book"
32 PRINT "* CAD Algorithms for Linear Microwave Circuits * publ
ished by
34 PRINT "Artech House,Inc..It includes twenty five computer pr
ograms "
```

```
36 PRINT "whose names and purpose are given in Tables 3.1, 3.2
and 3.3."
38 PRINT "The organization of this design kit is shown in Figur
e 3.1."
40 PRINT "According to this organization structure the initial
program"
42 PRINT "MENU allows us to select the type of the circuit to b
e designed."
44 PRINT "For this aim we have to call the suitable computer pr
ogram
46 PRINT "by entering its name."
48 PRINT
50 INPUT "Enter 'END' or the name of the required computer prog
ram:",A$
52 IF A$="PI-T" THEN  CHAIN "PI_T",200
54 IF A$="z-S-y" THEN  CHAIN "Z_S_Y",400
56 IF A$="A-S-T" THEN  CHAIN "A_S_T",700
58 IF A$="Z/L/S" THEN  CHAIN "Z_L_S",1100
60 IF A$="SMITH" THEN  CHAIN "SMITH",1400
62 IF A$="STL" THEN  CHAIN "STL",1600
64 IF A$="Q-IT" THEN  CHAIN "Q_IT",2500
66 IF A$="R-IT" THEN  CHAIN "R_IT",3000
68 IF A$="DC" THEN  CHAIN "DC",3500
70 IF A$="HPD" THEN  CHAIN "HPD",4400
72 IF A$="LPF" THEN  CHAIN "LPF",4800
74 IF A$="BPF" THEN  CHAIN "BPF",5300
76 IF A$="BSF" THEN  CHAIN "BSF",6200
78 IF A$="HF" THEN  CHAIN "HF",6600
80 IF A$="RA" THEN  CHAIN "RA",6800
82 IF A$="NMC" THEN  CHAIN "NMC",7000
84 IF A$="ECL" THEN  CHAIN "ECL",7700
86 IF A$="SL" THEN  CHAIN "SL",7900
88 IF A$="TL" THEN  CHAIN "TL",8200
90 IF A$="ML" THEN  CHAIN "ML",8400
92 IF A$="IML" THEN  CHAIN "IML",8700
94 IF A$="CSL" THEN  CHAIN "CSL",8900
96 IF A$="CTL" THEN  CHAIN "CTL",9300
98 IF A$="CML" THEN CHAIN "CML",9600
100 IF A$="END" THEN STOP
102 PRINT
104 PRINT "Wrong name !"
106 GOTO 48
```

REFERENCES

[1] Matthaei, G.L., L. Young, and E.M.T. Jones, *Microwave Filters, Impedance Matching Networks and Coupling Structures,* Norwood, MA: Artech House, 1980.

[2] Microwave integrated circuits. Special issue of *IEEE Trans. Microwave Theory and Techniques,* Vol. MTT-19, No. 7, July 1971.

[3] Howe, H., Jr., *Stripline Circuit Design,* Norwood, MA: Artech House, 1974.

[4] Frey, J., and K. Bhasin, eds., *Microwave Integrated Circuits,* 2nd Ed., Norwood, MA: Artech House, 1987.

[5] Chua, L.O., and P.M. Lin, *Computer Aided Analysis of Electronic Circuits. Algorithms and Computational Techniques,* Englewood Cliffs, NJ: Prentice-Hall, 1975.

[6] Allen, J.L., and M.W. Medley, Jr., *Microwave Circuit Design Using Programmable Calculators,* Norwood, MA: Artech House, 1980.

[7] Gupta, K.C., R. Garg, and R. Chadha, *Computer-Aided Design of Microwave Circuits,* Norwood, MA: Artech House, 1981.

[8] Edwards, T.C., *Foundations for Microstrip Circuit Design,* New York: John Wiley and Sons, 1982.

[9] Hoffmann, R.K., *Handbook of Microwave Integrated Circuits,* Norwood, MA: Artech House, 1987.

Chapter 4
ALGORITHMS FOR NETWORK ANALYSIS WITH ABCD-, Y-, Z-, T-, AND S-PARAMETERS

4.1 INTRODUCTION

Electrical networks with lumped and distributed parameters are most frequently characterized by *ABCD*-, *Y*-, *Z*-, *T*- and *S*-parameters. The first three sets of parameters relate the terminal voltages and currents, while the *T*- and *S*-parameters are closely related to the power incident on and reflected from the particular ports of the network under analysis (see [1–3]).

At microwave frequencies, the scattering matrix description is widely used because *S*-parameters have a simple and straightforward physical interpretation, and in most cases can be found experimentally without difficulty. The *S*-parameters, however, are unsuitable for analyzing cascaded networks. So, for this application, we convert them to *T*- or *ABCD*-parameters. In the case of a two-port network, the parameters can be converted by using the computer program A-S-T presented in Section 4.4. Similarly, for transformation of *S*-parameters to *Y*- or *Z*-parameters the expressions, and the computer program z-S-y related to them, described in Section 4.3, can be used. If the analyzed two-port network is characterized by *S*-parameters, for a given terminating impedance, the insertion loss and driving-point impedance at the input port can be calculated by using the program Z/L/S, included in Section 4.5.

Another fundamental problem, the transformation of a terminating complex impedance through a lossy multisection transmission line, is considered in Section 4.6. The computer program given in this section is named SMITH because it is based on the same impedance equation as the Smith chart.

Section 4.7 presents the algorithm and accompanying computer program STL, which make possible the design of a multisection stepped transmission line, the *R-transformer*. In the general case, the transmission line designed in this way can be used as a broadband impedance transformer, stopband filter, or impedance-transforming stopband filter. The well-known quarter-wavelength Chebyschev impedance transformers are simply particular cases of this generalized line.

4.2 TRANSFORMATIONS OF π TO T SECTION AND *VICE VERSA*

The electrical schemes of the π and T sections composed of lumped nonzero impedances are shown in Figure 4.1. Let us find the π section that is equivalent (at one frequency) to a given T section. We can do so by using the following $\pi(T)$ transformation:

$$
\begin{aligned}
Z_a &= N/Z_3 \\
Z_b &= N/Z_2 \\
Z_c &= N/Z_1
\end{aligned}
\tag{4.1}
$$

where $N = Z_1Z_2 + Z_2Z_3 + Z_1Z_3$.

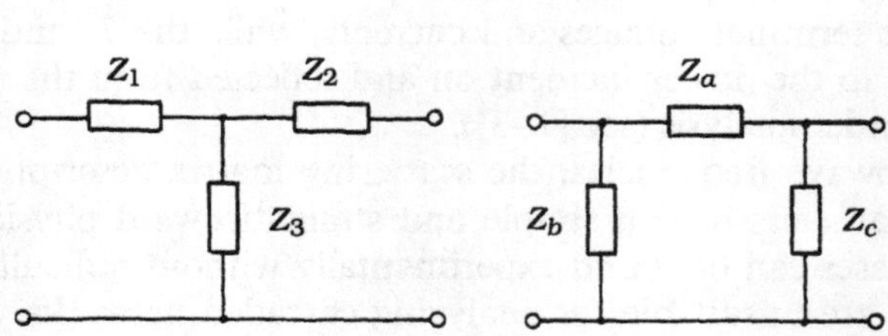

Figure 4.1 The lumped element sections, (a) of type T, (b) of type π.

Similarly, the $T(\pi)$ transformation is conducted in terms of the following simple relationships:

$$
\begin{aligned}
Z_1 &= Z_aZ_b/D \\
Z_2 &= Z_aZ_c/D \\
Z_3 &= Z_bZ_c/D
\end{aligned}
\tag{4.2}
$$

where $D = Z_a + Z_b + Z_c$ and $D \neq 0$.

The transformations given above can be realized automatically by using the computer program PI-T. In this program, the impedances of both transformed circuits are presented in exponential form, and the phase angles of these impedances are expressed in radians.

Check Calculations

The computer program PI-T can be verified by using the following data (see Figure 4.1).

Input Data

$T(\pi)$

Z_a = 300.00 exp (j 0.000) Ω

Z_b = 141.40 exp (j 0.785) Ω

Z_c = 223.60 exp (j 0.463) Ω

Results

Z_1 = 67.07 exp (j 0.463) Ω

Z_2 = 106.06 exp (j 0.141) Ω

Z_3 = 49.99 exp (j 0.926) Ω

Listing of Program PI-T

```
200 REM PI-T
202 CLS
204 DIM M(3)
206 DIM A(3)
208 PRINT "PI(T) AND T(PI) TRANSFORMATIONS"
210 PRINT
212 PRINT "Data:"
214 PRINT
216 INPUT "Enter the name ( T(PI) or PI(T) ) of the transformat
ion to be done:",P$
218 PRINT
220 LET PI=3.141593
222 IF P$="PI(T)" THEN  GOTO 232
224 IF P$="T(PI)" THEN  GOTO 264
226 PRINT "Wrong command !"
228 PRINT
230 GOTO 214
232 PRINT "Enter the impedances in  ohms:"
234 PRINT
236 INPUT ;"Z(1)=",M(1):INPUT ;" exp(j",A(1):PRINT ")"
238 INPUT ;"Z(2)=",M(2):INPUT ;" exp(j",A(2):PRINT ")"
240 INPUT ;"Z(3)=",M(3):INPUT ;" exp(j",A(3):PRINT ")"
242 PRINT
244 LET R=M(1)*M(2)* COS (A(1)+A(2))+M(2)*M(3)* COS (A(2)+A(3))
+M(1)*M(3)* COS (A(1)+A(3))
246 LET X=M(1)*M(2)* SIN (A(1)+A(2))+M(2)*M(3)* SIN (A(2)+A(3))
+M(1)*M(3)* SIN (A(1)+A(3))
248 GOSUB 294
250 PRINT
```

```
252 PRINT "Results, see Fig. 4.1(b)"
254 PRINT
256 PRINT "Za=";M1/M(3);" exp(j";V1-A(3);")"
258 PRINT "Zb=";M1/M(2);" exp(j";V1-A(2);")"
260 PRINT "Zc=";M1/M(1);" exp(j";V1-A(1);")"
262 GOTO 320
264 PRINT "Enter impedances in  ohms:"
266 PRINT
268 INPUT ;"Za=",M(1):INPUT ;" exp(j",A(1):PRINT ")"
270 INPUT ;"Zb=",M(2):INPUT ;" exp(j",A(2):PRINT ")"
272 INPUT ;"Zc=",M(3):INPUT ;" exp(j",A(3):PRINT ")"
274 PRINT
276 LET R=M(1)* COS (A(1))+M(2)* COS (A(2))+M(3)* COS (A(3))
278 LET X=M(1)* SIN (A(1))+M(2)* SIN (A(2))+M(3)* SIN (A(3))
280 GOSUB 294
282 PRINT "Results, see Fig. 4.1(a)"
284 PRINT
286 PRINT "Z1=";M(1)*M(2)/M1;" exp(j";A(1)+A(2)-V1;")"
288 PRINT "Z2=";M(1)*M(3)/M1;" exp(j";A(1)+A(3)-V1;")"
290 PRINT "Z3=";M(2)*M(3)/M1;" exp(j";A(2)+A(3)-V1;")"
292 GOTO 320
294 REM Subroutine 294
296 LET M1= SQR (R*R+X*X)
298 IF M1=0 THEN  LET V1=0: RETURN
300 IF (X/M1) >= .999999 THEN  LET V1= PI /2: RETURN
302 IF (X/M1) <= -.999999 THEN  LET V1=- PI /2: RETURN
304 IF  ABS (R) <= 1E-09 THEN  GOTO 312
306 IF R<0 THEN  GOTO 310
308 LET V1= ATN (X/R): RETURN
310 LET V1= PI + ATN (X/R): RETURN
312 IF X=0 AND R>0 THEN  LET V1=0: RETURN
314 IF X=0 AND R<0 THEN  LET V1= PI : RETURN
316 IF X>0 THEN  LET V1= PI /2- ATN (R/X): RETURN
318 LET V1=- PI /2+ ATN (R/X): RETURN
320 REM Chain subroutine
322 PRINT
324 INPUT"Enter 'CONT':",N$
326 IF N$="CONT" THEN CLS:CHAIN "MENU",10
328 GOTO 322
```

4.3 RELATIONSHIPS BETWEEN *S*-, *Y*-, AND *Z*-MATRICES OF A TWO-PORT NETWORK

The *Y*-, *Z*-, and *S*-parameters are most frequently used in analysis and synthesis of two-port networks. The first two sets of parameters are related to the terminal voltages and currents in a different way, while the *S*-parameters are closely related to the power incident on and reflected from the ports of the network being described.

For the electrical network shown in Figure 4.2, the parameters are defined as [2,3]:

$$\begin{bmatrix} U_1 \\ U_2 \end{bmatrix} = Z_0 \begin{bmatrix} z_{11}, z_{12} \\ z_{21}, z_{22} \end{bmatrix} \begin{bmatrix} I_1 \\ I_2 \end{bmatrix}$$

$$\begin{bmatrix} I_1 \\ I_2 \end{bmatrix} = \frac{1}{Z_0} \begin{bmatrix} y_{11}, y_{12} \\ y_{21}, y_{22} \end{bmatrix} \begin{bmatrix} U_1 \\ U_2 \end{bmatrix} \quad (4.3)$$

$$\begin{bmatrix} b_1 \\ b_2 \end{bmatrix} = \begin{bmatrix} S_{11}, S_{12} \\ S_{21}, S_{22} \end{bmatrix} \begin{bmatrix} a_1 \\ a_2 \end{bmatrix} \text{ at } Z_0$$

where

$$\begin{aligned} a_1 &= (U_1 + I_1 Z_0)/(2\sqrt{Z_0}) \\ a_2 &= (U_2 + I_2 Z_0)/(2\sqrt{Z_0}) \\ b_1 &= (U_1 - I_1 Z_0)/(2\sqrt{Z_0}) \\ b_2 &= (U_2 - I_2 Z_0)/(2\sqrt{Z_0}) \end{aligned} \quad (4.4)$$

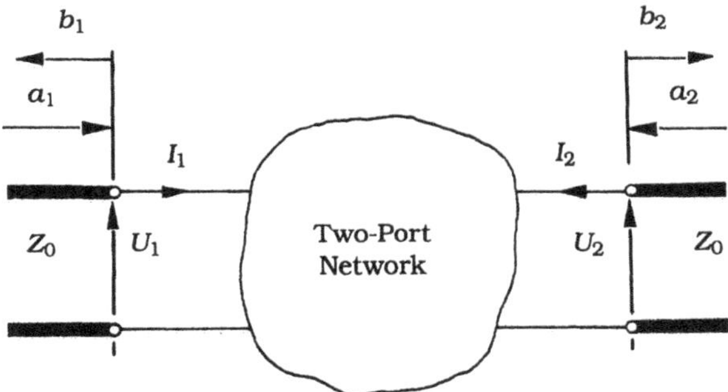

Figure 4.2 The voltages, currents and scattering waves relevant to the definition of matrices z, y and S or a two-port network.

Each of previously defined matrices can be expressed in terms of the other (see Figure 4.3) by using the general transformation $V(W)$ defined as:

$$v_{11} = \frac{[(k_1 + k_2 w_{11})(k_1 - k_2 w_{22}) + w_{12} w_{21}]k_3 + (1 - k_1)w_{22}}{D}$$

$$v_{12} = \frac{w_{12}}{D} k_4 \quad (4.5)$$

$$v_{21} = \frac{w_{21}}{D} k_4$$

$$v_{22} = \frac{[(k_1 + k_2 w_{22})(k_1 - k_2 w_{11}) + w_{12} w_{21}] k_3 + (1 - k_1) w_{11}}{D}$$

where $D = (k_1 + k_5 w_{11})(k_1 + k_5 w_{22}) - w_{12} w_{21}$ and $D \neq 0$.

The values of coefficients k_i, where $i = 1, 2, \ldots, 5$, for the individual transformations are given in Table 4.1.

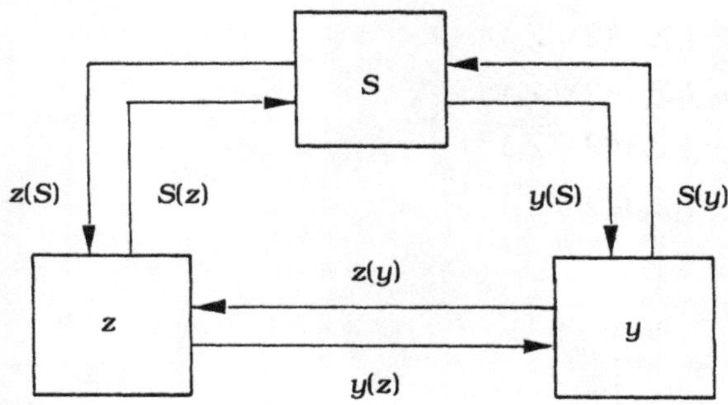

Figure 4.3 The schematic diagram of the reciprocal conversions between matrices z, y and S of a two-port network.

Table 4.1
The Coefficients of the General Transformation (4.5)

Transformations	k_1	k_2	k_3	k_4	k_5
$S(z)$	1	−1	−1	2	1
$S(y)$	1	−1	1	−2	1
$z(S)$	1	1	1	2	−1
$y(S)$	1	−1	1	−2	1
$z(y)$	0	0	0	−1	1
$y(z)$	0	0	0	−1	1

As an example, let us assume that the admittance matrix $[y]$ of a two-port network is known. In this case, the corresponding scattering matrix $[S]$ can be calculated by using the set of formulas (4.5) at $k_1 = 1$, $k_2 = -1$, $k_3 = 1$, $k_4 = -2$, and $k_5 = 1$. Note that the S-matrix obtained in this way is valid only at reference impedance Z_0, which is determined by the set of relations (4.3).

All of the mathematical relations presented above are used in the computer program z-S-y. In this program, the complex elements of the matrices being transformed are in exponential form, and the phase angles of these elements are given in radians. Output results are presented in the same form.

Check Calculations

To verify the computer program z-S-y, we can use the following sample calculations.

(1) Input Data

$y(S)$

$S_{11} = 0.600 \exp(j\ 0.100)$
$S_{12} = 0.800 \exp(j\ 0.000)$
$S_{21} = 2.000 \exp(j\ 1.000)$
$S_{22} = 0.600 \exp(j\ 0.100)$

Results

$y_{11} = 0.969 \exp(j\ 1.302)$
$y_{12} = 0.784 \exp(j\ 3.743)$
$y_{21} = 1.960 \exp(-j\ 1.539)$
$y_{22} = 0.969 \exp(j\ 1.302)$

(2) Input Data

$z(S)$

$S_{11} = 0.600 \exp(j\ 0.100)$
$S_{12} = 0.800 \exp(j\ 0.000)$
$S_{21} = 2.000 \exp(j\ 1.000)$
$S_{22} = 0.600 \exp(j\ 0.100)$

Results

$z_{11} = 1.265 \exp(j\ 2.739)$
$z_{12} = 1.023 \exp(j\ 2.039)$
$z_{21} = 2.559 \exp(j\ 3.039)$
$z_{22} = 1.265 \exp(j\ 2.739)$

Listing of program z-S-y

```
400 REM z-S-y
402 CLS
404 PRINT "TRANSFORMATIONS:S(z), z(S), S(y), y(S), z(y) AND y(z
) "
406 PRINT
408 LET Q=1
410 DIM K(5)
412 DIM M(2,2)
414 DIM V(2,2)
416 PRINT "Data:"
418 PRINT
420 INPUT "Enter the name of the transformation to be done:",P$
422 FOR I=1 TO 5
424 LET K(I)=1
426 NEXT I
428 LET PI=3.141593
430 IF P$="S(z)" THEN  LET Q=0: GOTO 446
432 IF P$="z(S)" THEN  LET Q=0: GOTO 454
434 IF P$="S(y)" THEN  GOTO 460
436 IF P$="y(S)" THEN  GOTO 460
438 IF P$="z(y)" THEN  GOTO 466
440 IF P$="y(z)" THEN  GOTO 466
442 PRINT
444 PRINT "Wrong command !": GOTO 418
446 LET K(2)=-1
448 LET K(3)=-1
450 LET K(4)=2
```

```
452 GOTO 474
454 LET K(4)=2
456 LET K(5)=-1
458 GOTO 474
460 LET K(2)=-1
462 LET K(4)=-2
464 GOTO 474
466 LET K(1)=0
468 LET K(2)=0
470 LET K(3)=0
472 LET K(4)=-1
474 LET R$=MID$(P$,3,1)
476 LET Q$=MID$(P$,1,1)
478 PRINT
480 FOR I=1 TO 2
482 FOR J=1 TO 2
484 PRINT USING "&(# #)=";R$;I;J;:INPUT ;"",M(I,J):INPUT ;" exp
(j", V(I,J):PRINT")"
486 NEXT J
488 NEXT I
490 LET R=K(1)*K(1)+K(1)*K(5)*M(1,1)* COS(V(1,1))+K(1)*K(5)*M(2
,2)* COS(V(2,2))+K(5)*K(5)*M(1,1)*M(2,2)* COS (V(1,1)+V(2,2))-M
(1,2)*M(2,1)* COS (V(1,2)+V(2,1))
492 LET X=K(1)*K(5)*M(1,1)* SIN(V(1,1))+K(1)*K(5)*M(2,2)* SIN(V
(2,2))+K(5)*K(5)*M(1,1)*M(2,2)* SIN (V(1,1)+V(2,2))-M(1,2)*M(2,
1)* SIN (V(1,2)+V(2,1))
494 GOSUB 528
496 LET MD=M
498 LET VD=V
500 LET R=K(3)*(K(1)*K(1)+K(1)*K(2)*M(1,1)* COS ( V(1,1))-K(1)*
K(2)*M(2,2)* COS(V(2,2))-K(2)*K(2)*M(1,1)*M(2,2)* COS (V(1,1)+V
(2,2))+M(1,2)*M(2,1)* COS (V(1,2)+V(2,1)))+(1-K(1))*M(2,2)* COS
( V(2,2))
502 LET X=K(3)*(K(1)*K(2)*M(1,1)* SIN(V(1,1))-K(1)*K(2)*M(2,2)*
 SIN(V(2,2))-K(2)*K(2)*M(1,1)*M(2,2)* SIN (V(1,1)+V(2,2))+M(1,2
)*M(2,1)* SIN (V(1,2)+V(2,1)))+(1-K(1))*M(2,2)* SIN(V(2,2))
504 PRINT
506 PRINT "Results, see Fig. 4.3"
508 PRINT
510 GOSUB 528
512 PRINT Q$;"(11)=";M/MD;"exp(j";V-VD;")"
514 PRINT Q$;"(12)="; ABS (K(4))*M(1,2)/MD;"exp(j"; PI *Q+V(1,2
)-VD;")"
516 PRINT Q$;"(21)="; ABS (K(4))*M(2,1)/MD;"exp(j"; PI *Q+V(2,1
)-VD;")"
518 LET R=K(3)*(K(1)*K(1)+K(1)*K(2)*M(2,2)* COS(V(2,2))-K(1)*K(
2)*M(1,1)* COS(V(1,1))-K(2)*K(2)*M(1,1)*M(2,2)* COS (V(1,1)+V(2
,2))+M(1,2)*M(2,1)* COS (V(1,2)+V(2,1)))+(1-K(1))*M(1,1)* COS(V
(1,1))
520 LET X=K(3)*(K(1)*K(2)*M(2,2)* SIN(V(2,2))-K(1)*K(2)*M(1,1)*
 SIN(V(1,1))-K(2)*K(2)*M(1,1)*M(2,2)* SIN (V(1,1)+V(2,2))+M(1,2
)*M(2,1)* SIN (V(1,2)+V(2,1)))+(1-K(1))*M(1,1)* SIN(V(1,1))
522 GOSUB 528
524 PRINT Q$;"(22)=";M/MD;"exp(j";V-VD;")"
526 GOTO 554
528 REM Subroutine 528
```

```
530 LET M= SQR (R*R+X*X)
532 IF M=0 THEN  LET V=0: RETURN
534 IF (X/M) >= .999999 THEN  LET V= PI /2: RETURN
536 IF (X/M) <= -.999999 THEN  LET V=- PI /2: RETURN
538 IF  ABS (R) <= 1E-09 THEN  GOTO 546
540 IF R<0 THEN  GOTO 544
542 LET V= ATN (X/R): RETURN
544 LET V= PI + ATN (X/R): RETURN
546 IF X=0 AND R>0 THEN  LET V=0: RETURN
548 IF X=0 AND R<0 THEN  LET V= PI : RETURN
550 IF X>0 THEN  LET V= PI /2- ATN (R/X): RETURN
552 LET V=- PI /2+ ATN (R/X): RETURN
554 REM Chain subroutine
556 PRINT
558 INPUT"Enter 'CONT':",N$
560 IF N$="CONT" THEN CLS:CHAIN "MENU",10
562 GOTO 556
```

4.4 RELATIONSHIPS BETWEEN THE *ABCD*, *S*- AND *T*-MATRICES OF A TWO-PORT NETWORK

The electrical network we will now consider is shown in Figure 4.4, and has input port 1 and output port 2. The quantities a_i and b_i (see (4.4)), represent the power waves incident on and reflected from port i, where $i = 1$ or 2. According to [3,5], the standard relations between these quantities are:

$$\begin{aligned} b_1 &= S_{11}a_1 + S_{12}a_2 \\ b_2 &= S_{21}a_1 + S_{22}a_2 \end{aligned} \tag{4.6}$$

and

$$\begin{aligned} b_1 &= T_{11}a_2 + T_{12}b_2 \\ a_1 &= T_{21}a_2 + T_{22}b_2 \end{aligned} \tag{4.7}$$

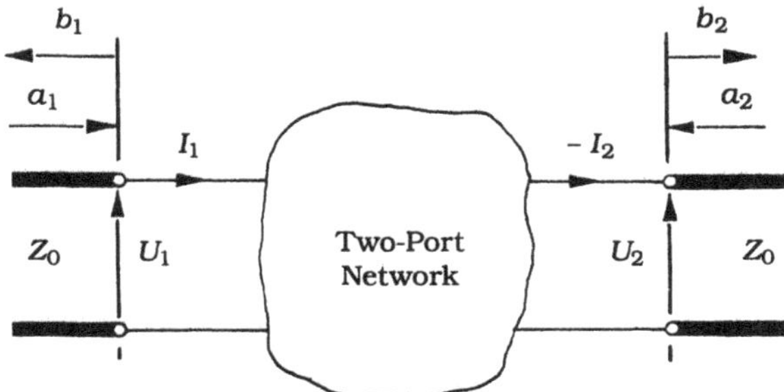

Figure 4.4 The voltages, currents and scattering waves relevant to the definition of matrices *ABCD* and *S* of a two-port network.

The elements of the transfer scattering matrix $[T]$, defined by (4.7), are related to the S-parameters as follows:

$$\begin{bmatrix} T_{11}, & T_{12} \\ T_{21}, & T_{22} \end{bmatrix} = \frac{1}{S_{21}} \begin{bmatrix} -\det[S], & S_{11} \\ -S_{22}, & 1 \end{bmatrix} \tag{4.8}$$

where $\det[S] = S_{11}S_{22} - S_{12}S_{21}$.

Similarly, the scattering matrix $[S]$ of the network under discussion can be obtained from the T-matrix by using the following relations:

$$\begin{bmatrix} S_{11}, & S_{12} \\ S_{21}, & S_{22} \end{bmatrix} = \frac{1}{T_{22}} \begin{bmatrix} T_{12}, & \det[T] \\ 1, & -T_{21} \end{bmatrix} \tag{4.9}$$

where $\det[T] = T_{11}T_{22} - T_{12}T_{21}$.

To convert the S-matrix into the T-matrix, and *vice versa,* the elements S_{21} and T_{22} should be nonzero quantities. If S_{21} (or T_{22}) is equal to zero, the input and output ports are isolated from each other. Note that T_{22} is the reciprocal of the forward transmission coefficient S_{21}, and for "normal" two-port networks T_{22} takes nonzero values. If element S_{21} of the S-matrix is equal to zero, the T-matrix corresponding to the S-matrix cannot be evaluated.

The $ABCD$-parameters for the two-port network, shown in Figure 4.4, are defined as follows:

$$\begin{aligned} U_1 &= AU_2 - BI_2 \\ I_1 &= CU_2 - DI_2 \end{aligned} \tag{4.10}$$

Some useful $ABCD$-parameter relationships for two-port networks are given in [4,5,8]. Therefore, here we will only show how these parameters are related to the S-parameters defined by (4.6) and *vice versa.* The $ABCD$-matrix can be converted into the S-matrix by using the following relations:

$$\begin{aligned} A &= (1 + S_{11} - S_{22} - \det[S])/(2S_{21}) \\ B &= (1 + S_{11} + S_{22} + \det[S])Z_0/(2S_{21}) \\ C &= (1 - S_{11} - S_{22} + \det[S])/(2Z_0S_{21}) \\ D &= (1 - S_{11} + S_{22} - \det[S])/(2S_{21}) \end{aligned} \tag{4.11}$$

From these relations, for a network with its S_{21} parameter equal to zero, the $ABCD$-parameters hence cannot be calculated. The reverse conversion, from the $ABCD$-matrix into the S-matrix can be made by using the formulas:

$$S_{11} = (AZ_0 + B - CZ_0^2 - DZ_0)/P$$
$$S_{12} = 2Z_0(AD - BC)/P \qquad (4.12)$$
$$S_{21} = 2Z_0/P$$
$$S_{22} = (-AZ_0 + B - CZ_0^2 + DZ_0)/P$$

where $P = AZ_0 + B + CZ_0^2 + DZ_0$ and $P \neq 0$.

For the reciprocal two-port networks, the expression $(AD - BC)$ is equal to unity, and therefore parameters S_{12} and S_{21} are equal.

The above relationships constitute a mathematical basis of the computer program A-S-T. This program allows us to calculate the elements of the *T*-matrix of a two-port network when its *S*-matrix is known. The reverse transformation from the *T*-matrix into the *S*-matrix can be done in like manner. Similar transformations also can be made between *S*- and *ABCD*-matrices. All complex elements of the matrices being converted should be given in exponential form, and the phase angles of these elements should be expressed in radians. Output results are obtained in the same form.

Check Calculations

In the computer program, A-S-T, the above transformations (4.8), (4.9), (4.11), and (4.12) are denoted as $T(S)$, $S(T)$, $A(S)$, and $S(A)$, respectively.

Some typical results of calculations obtained by using this program are presented below.

1 Input Data

$S(A)$

$Z_0 = 50$, ohm
$A = 1.000 \exp(j\ 0.000)$
$B = 100 \exp(j\pi/2)$
$C = 0.000 \exp(j\ 0.000)$
$D = 1.000 \exp(j\ 0.000)$

Results

$S_{11} = 0.707 \exp(j\ 0.785)$
$S_{12} = 0.707 \exp(-j\ 0.785)$
$S_{21} = 0.707 \exp(-j\ 0.785)$
$S_{22} = 0.707 \exp(j\ 0.785)$

2 Input Data

$T(S)$

$S_{11} = 0.277 \exp(-j\ 1.029)$
$S_{12} = 0.078 \exp(j\ 1.623)$
$S_{21} = 1.920 \exp(j\ 1.117)$
$S_{22} = 0.484 \exp(-j\ 0.541)$

Results

$T_{11} = 0.123 \exp(j\ 1.075)$
$T_{12} = 0.144 \exp(-j\ 2.146)$
$T_{21} = 0.252 \exp(j\ 1.483)$
$T_{22} = 0.520 \exp(-j\ 1.117)$

Listing of Programs A-S-T

```
700 REM A-S-T
702 CLS
704 PRINT "TRANSFORMATIONS:S(T), T(S), A(S) AND S(A)"
```

```
706 PRINT
708 PRINT "Data:"
710 PRINT
712 INPUT "Enter the name of transformation to be done:",P$
714 PRINT
716 LET PI= 3.141593
718 IF P$="S(T)" THEN  GOTO 728
720 IF P$="T(S)" THEN  GOTO 758
722 IF P$="A(S)" THEN  GOTO 788
724 IF P$="S(A)" THEN  GOTO 852
726 PRINT "Wrong command !": GOTO 710
728 REM S(T) transformation
730 INPUT ;"T(11)=",M11:INPUT ;" exp(j",V11: PRINT ")"
732 INPUT ;"T(12)=",M12:INPUT ;" exp(j",V12: PRINT ")"
734 INPUT ;"T(21)=",M21:INPUT ;" exp(j",V21: PRINT ")"
736 INPUT ;"T(22)=",M22:INPUT ;" exp(j",V22: PRINT ")"
738 PRINT
740 PRINT "Results:"
742 PRINT
744 GOSUB 924
746 GOSUB 932
748 PRINT "S(11)=";M12/M22;"exp(j";V12-V22;")"
750 PRINT "S(12)=";M/M22;"exp(j";V-V22;")"
752 PRINT "S(21)=";1/M22;"exp(j";-V22;")"
754 PRINT "S(22)=";M21/M22;"exp(j"; PI +V21-V22;")"
756 GOTO 958
758 REM T(S) transformation
760 INPUT ;"S(11)=",M11:INPUT ;" exp(j",V11: PRINT ")"
762 INPUT ;"S(12)=",M12:INPUT ;" exp(j",V12: PRINT ")"
764 INPUT ;"S(21)=",M21:INPUT ;" exp(j",V21: PRINT ")"
766 INPUT ;"S(22)=",M22:INPUT ;" exp(j",V22: PRINT ")"
768 PRINT
770 PRINT "Results:"
772 PRINT
774 GOSUB 924
776 GOSUB 932
778 PRINT "T(11)=";M/M21;"exp(j"; PI +V-V21;")"
780 PRINT "T(12)=";M11/M21;"exp(j";V11-V21;")"
782 PRINT "T(21)=";M22/M21;"exp(j"; PI +V22-V21;")"
784 PRINT "T(22)=";1/M21;"exp(j";-V21;")"
786 GOTO 958
788 REM A(S) transformation
790 INPUT "Zo=",ZO
792 IF ZO>0 THEN  GOTO 796
794 PRINT "Error:Zo <= 0": GOTO 790
796 PRINT
798 INPUT ;"S(11)=",M11:INPUT ;" exp(j",V11: PRINT ")"
800 INPUT ;"S(12)=",M12:INPUT ;" exp(j",V12: PRINT ")"
802 INPUT ;"S(21)=",M21:INPUT ;" exp(j",V21: PRINT ")"
804 INPUT ;"S(22)=",M22:INPUT ;" exp(j",V22: PRINT ")"
806 PRINT
808 PRINT "Results:"
810 PRINT
812 GOSUB 924
814 LET RD=R
```

```
816 LET XD=X
818 LET R=1+M11* COS(V11)-M22* COS(V22)-RD
820 LET X=M11* SIN(V11)-M22* SIN(V22)-XD
822 GOSUB 932
824 PRINT "A=";M/(2*M21);"exp(j";V-V21;")"
826 LET R=1+M11* COS(V11)+M22* COS(V22)+RD
828 LET X=M11* SIN(V11)+M22* SIN(V22)+XD
830 GOSUB 932
832 PRINT "B=";M*ZO/(2*M21);"exp(j";V-V21;")"
834 LET R=1-M11* COS(V11)-M22* COS(V22)+RD
836 LET X=-M11* SIN(V11)-M22* SIN(V22)+XD
838 GOSUB 932
840 PRINT "C=";M/(2*ZO*M21);"exp(j";V-V21;")"
842 LET R=1-M11* COS(V11)+M22* COS(V22)-RD
844 LET X=-M11* SIN(V11)+M22* SIN(V22)-XD
846 GOSUB 932
848 PRINT "D=";M/(2*M21);"exp(j";V-V21;")"
850 GOTO 958
852 REM S(A) transformation
854 INPUT "Zo=",ZO
856 IF ZO>0 THEN  GOTO 860
858 PRINT "Error:Zo <= 0": GOTO 854
860 PRINT
862 INPUT ;"A=",M11:INPUT ;" exp(j",V11: PRINT ")"
864 INPUT ;"B=",M12:INPUT ;" exp(j",V12: PRINT ")"
866 INPUT ;"C=",M21:INPUT ;" exp(j",V21: PRINT ")"
868 INPUT ;"D=",M22:INPUT ;" exp(j",V22: PRINT ")"
870 PRINT
872 PRINT "Results:"
874 PRINT
876 GOSUB 924
878 LET RD=R
880 LET XD=X
882 GOSUB 932
884 LET MD=M
886 LET VD=V
888 LET R=M11*ZO* COS(V11)+M12* COS(V12)+ZO*ZO*M21* COS(V21)+ZO
*M22* COS(V22)
890 LET X=M11*ZO* SIN(V11)+M12* SIN(V12)+ZO*ZO*M21* SIN(V21)+ZO
*M22* SIN(V22)
892 GOSUB 932
894 LET MP=M
896 LET VP=V
898 IF MP>0 THEN  GOTO 902
900 PRINT "The transformation cannot be done since P=0, see (4.
12)": GOTO 958
902 LET R=ZO*M11* COS(V11)+M12* COS(V12)-ZO*ZO*M21* COS(V21)-ZO
*M22* COS(V22)
904 LET X=ZO*M11* SIN(V11)+M12* SIN(V12)-ZO*ZO*M21* SIN(V21)-ZO
*M22* SIN(V22)
906 GOSUB 932
908 PRINT "S(11)=";M/MP;"exp(j";V-VP;")"
910 PRINT "S(12)=";2*ZO*MD/MP;"exp(j";VD-VP;")"
912 PRINT "S(21)=";2*ZO/MP;"exp(j";-VP;")"
914 LET R=-ZO*M11* COS(V11)+M12* COS(V12)-ZO*ZO*M21* COS(V21)+Z
```

```
O*M22* COS(V22)
916 LET X=-ZO*M11* SIN(V11)+M12* SIN(V12)-ZO*ZO*M21* SIN(V21)+Z
O*M22* SIN(V22)
918 GOSUB 932
920 PRINT "S(22)=";M/MP;"exp(j";V-VP;")"
922 GOTO 958
924 REM Subroutine 924
926 LET R=M11*M22* COS (V11+V22)-M12*M21* COS (V12+V21)
928 LET X=M11*M22* SIN (V11+V22)-M12*M21* SIN (V12+V21)
930 RETURN
932 REM Subroutine 932
934 LET M= SQR (R*R+X*X)
936 IF M=0 THEN  LET V=0: RETURN
938 IF (X/M) >= .999999 THEN  LET V= PI /2: RETURN
940 IF (X/M) <= -.999999 THEN  LET V=- PI /2: RETURN
942 IF  ABS (R) <= 1E-09 THEN  GOTO 950
944 IF R<0 THEN  GOTO 948
946 LET V= ATN (X/R): RETURN
948 LET V= PI + ATN (X/R): RETURN
950 IF X=0 AND R>0 THEN  LET V=0: RETURN
952 IF X=0 AND R<0 THEN  LET V= PI  : RETURN
954 IF X>0 THEN  LET V= PI /2- ATN (R/X): RETURN
956 LET V=- PI /2+ ATN (R/X): RETURN
958 REM Chain subroutine
960 PRINT
962 INPUT"Enter 'CONT':",N$
964 IF N$="CONT" THEN CLS:CHAIN "MENU",10
966 GOTO 960
```

4.5 S-PARAMETER EXPRESSIONS USEFUL FOR THE ANALYSIS OF TWO-PORT NETWORKS

S-parameters are frequently used to analyze two-port networks, as shown in Figure 4.5. Let us therefore assume that this network is characterized by the *S*-parameters evaluated at a reference impedance equal to Z_0.

We can easily check that the driving point impedance at the input port 1 is given by:

$$Z_{in} = Z_0(1 + \Gamma_{in})/(1 - \Gamma_{in}) \tag{4.13}$$

where

$$\Gamma_{in} = S_{11} + S_{12}S_{21}\Gamma_l/(1 - S_{22}\Gamma_l),$$
$$\Gamma_l = (Z_l - Z_0)/(Z_l + Z_0)$$

and Z_l is an arbitrary complex load impedance.

The *voltage standing wave ratio* (VSWR) at the input port 1 is, therefore:

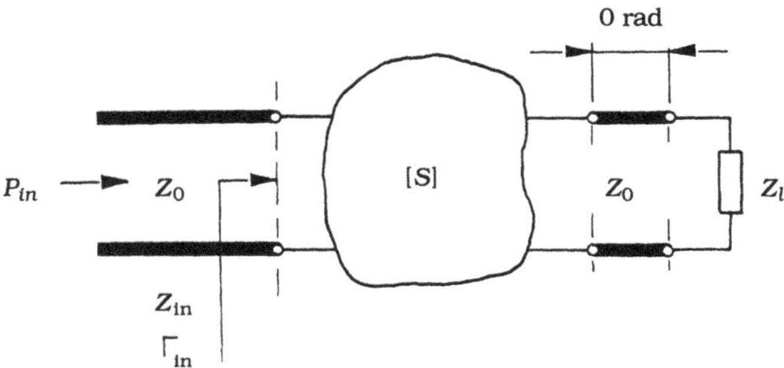

Figure 4.5 Transformation of the terminating impedance via a two-port network characterized by S-parameters.

$$\text{VSWR} = (1 + |\Gamma_{in}|)/(1 - |\Gamma_{in}|) \tag{4.14}$$

For further consideration, we need to explain the physical meanings of the input power P_{in} and the load power P_l. By input power, P_{in}, we understand the power incident on port 1 (see Figure 4.5). The load power P_l is the difference of the powers incident on and reflected from the load impedance Z_l. The power P_l is often determined as the power delivered to the load [3]. Hence, the insertion loss coefficient L in dB, defined as the ratio of the input power P_{in} to the load power P_l, can be calculated by using the formula:

$$L = 10 \log \left[\frac{|1 - S_{22}\Gamma_1|^2}{|S_{21}|^2(1 - |\Gamma_1|^2)} \right], \text{dB} \tag{4.15}$$

In many practical cases, we need to calculate the scattering matrix $[S]$ for a network consisting of two-port circuits cascaded in a chain. An example of such a network is shown in Figure 4.6. The scattering matrix $[S]$ for the overall network under consideration is given in terms of the individual scattering matrices $[S_A]$ and $[S_B]$ by the set of equations [5,6]:

$$\begin{aligned} S_{11} &= (S_{11}^A + S_{12}^A S_{21}^A S_{11}^B)/D \\ S_{12} &= (S_{12}^A S_{12}^B)/D \\ S_{21} &= (S_{21}^A S_{21}^B)/D \\ S_{22} &= (S_{22}^B + S_{12}^B S_{21}^B S_{22}^A)/D \end{aligned} \tag{4.16}$$

where

$$D = 1 - S_{11}^B S_{22}^A$$

$$[S_A] = \begin{bmatrix} S_{11}^A, & S_{12}^A \\ S_{21}^A, & S_{22}^A \end{bmatrix} \text{ at } Z_0$$

$$[S_B] = \begin{bmatrix} S_{11}^B, & S_{12}^B \\ S_{21}^B, & S_{22}^B \end{bmatrix} \text{ at } Z_0$$

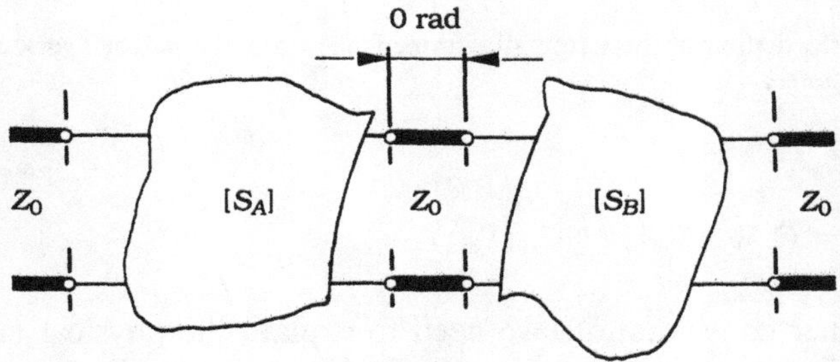

Figure 4.6 Two cascaded two-port networks characterized by S-parameters.

The S-matrix obtained in this way is valid only if the reference impedance is equal to Z_0.

Formulas (4.13) to (4.16) have been used in the computer program Z/L/S. To calculate the input impedance Z_{in}, VSWR, or insertion loss coefficient L, we first call its procedure Z-L, and then introduce into the computer memory the values of impedances Z_0, Z_l and the S-parameters. The second procedure of this program, denoted as S-S, makes possible the calculation of the S-matrix for the overall network shown in Figure 4.6. In this case, we introduce into the computer memory the elements of the scattering matrices $[S_A]$ and $[S_B]$.

In the program Z/L/S, all complex quantities (i.e., impedance Z_l and S-parameters) are presented in exponential form and their phase angles are expressed in radians.

Check Calculations

The computer program Z/L/S can be verified by using the following sample calculations:

(1) Input Data

Z-L

Z_0 = 50, Ω

S_{11} = 0.832 exp (j 0.588)

S_{12} = 0.554 exp (−j 0.982)

S_{21} = 0.554 exp (−j 0.982)

S_{22} = 0.832 exp (j 0.588)

Z_l = 75 exp (j 0.000), Ω

Results

Z_{in} = 167.715 exp (j 1.107), Ω

VSWR = 8.048

L = 4.062, dB

(2) Input Data

S-S

S^A_{11} = 0.832 exp (j 0.588)

S^A_{12} = 0.554 exp (−j 0.982)

S^A_{21} = 0.554 exp (−j 0.982)

S^A_{22} = 0.832 exp (j 0.588)

S^B_{11} = 0.447 exp (−j 1.107)

S^B_{12} = 0.894 exp (j 0.463)

S^B_{21} = 0.894 exp (j 0.463)

S^B_{22} = 0.447 exp (−j 1.107)

Results

S_{11} = 0.707 exp (j 0.784)

S_{12} = 0.705 exp (−j 0.785)

S_{21} = 0.705 exp (−j 0.785)

S_{22} = 0.707 exp (j 0.784)

Listing of Program Z/L/S

```
1100 REM Z/L/S
1102 CLS
1104 PRINT "Z-L AND S-S PROCEDURES"
1106 PRINT
1108 PRINT "Data:"
1110 PRINT
1112 INPUT "Enter the name of the procedure:",A$
1114 PRINT
1116 LET PI=3.141593
1118 IF A$="Z-L" THEN  GOTO 1124
1120 IF A$="S-S" THEN  GOTO 1218
1122 PRINT "Wrong command !": GOTO 1110
1124 REM Procedure Z-L
1126 INPUT "Zo=",ZO
1128 IF ZO <= 0 THEN  PRINT "Error:Zo <= 0": GOTO 1124
1130 PRINT
1132 INPUT ;"S(11)=",M11:INPUT ;" exp(j",V11:PRINT ")"
1134 INPUT ;"S(12)=",M12:INPUT ;" exp(j",V12:PRINT ")"
1136 INPUT ;"S(21)=",M21:INPUT ;" exp(j",V21:PRINT ")"
1138 INPUT ;"S(22)=",M22:INPUT ;" exp(j",V22:PRINT ")"
1140 PRINT
1142 INPUT ;"Zl=",ML:INPUT ;" exp(j",VL:PRINT ")"
1144 IF ML<0 THEN  PRINT "Error:ABS(Zl)<0": GOTO 1142
1146 PRINT
1148 LET R=ML* COS(VL)-ZO
1150 LET X=ML* SIN(VL)
1152 GOSUB 1288
```

```
1154 LET MN=M
1156 LET VN=V
1158 LET R=ML* COS(VL)+ZO
1160 GOSUB 1288
1162 LET MG=MN/M
1164 LET VG=VN-V
1166 LET R=1-M22*MG* COS (V22+VG)
1168 LET X=-M22*MG* SIN (V22+VG)
1170 GOSUB 1288
1172 LET MN=M12*M21*MG/M
1174 LET VN=V12+V21+VG-V
1176 LET GR=M11* COS(V11)+MN* COS(VN)
1178 LET GX=M11* SIN(V11)+MN* SIN(VN)
1180 LET R=1+GR
1182 LET X=GX
1184 GOSUB 1288
1186 LET MN=M
1188 LET VN=V
1190 LET R=1-GR
1192 LET X=-GX
1194 GOSUB 1288
1196 PRINT "Results:"
1198 PRINT
1200 PRINT "Zin=";ZO*MN/M;"exp(j";VN-V;")"
1202 LET MN= SQR (GR*GR+GX*GX)
1204 PRINT "VSWR=";(1+MN)/(1-MN)
1206 LET R=1-M22*MG* COS (V22+VG)
1208 LET X=-M22*MG* SIN (V22+VG)
1210 LET MN=R*R+X*X
1212 LET L=MN/(M21*M21*(1-MG*MG))
1214 PRINT "L[dB]=";10/ LOG(10)* LOG(L)
1216 GOTO 1314
1218 REM Procedure S-S
1220 PRINT "Enter S-parameters of the first two-port network, s
ee Fig. 4.6"
1222 PRINT
1224 INPUT ;"S(11)=",P11:INPUT ;" exp(j",V11: PRINT ")"
1226 INPUT ;"S(12)=",P12:INPUT ;" exp(j",V12: PRINT ")"
1228 INPUT ;"S(21)=",P21:INPUT ;" exp(j",V21: PRINT ")"
1230 INPUT ;"S(22)=",P22:INPUT ;" exp(j",V22: PRINT ")"
1232 PRINT
1234 PRINT "Enter S-parameters of the second two-port network"
1236 PRINT
1238 INPUT ;"S(11)=",Q11:INPUT ;" exp(j",W11: PRINT ")"
1240 INPUT ;"S(12)=",Q12:INPUT ;" exp(j",W12: PRINT ")"
1242 INPUT ;"S(21)=",Q21:INPUT ;" exp(j",W21: PRINT ")"
1244 INPUT ;"S(22)=",Q22:INPUT ;" exp(j",W22: PRINT ")"
1246 PRINT
1248 PRINT "Results:"
1250 PRINT
1252 LET R=1-Q11*P22* COS (W11+V22)
1254 LET X=-Q11*P22* SIN (W11+V22)
1256 GOSUB 1288
1258 LET MD=M
1260 LET VD=V
```

```
1262 IF MD>0 THEN  GOTO 1266
1264 PRINT "D=0, see (4.16)": GOTO 1314
1266 LET R=P11* COS(V11)+P12*P21*Q11/MD* COS (V12+V21+W11-VD)
1268 LET X=P11* SIN (V11)+P12*P21*Q11/MD* SIN (V12+V21+W11-VD)
1270 GOSUB 1288
1272 PRINT "S(11)=";M;"exp(j";V;")"
1274 PRINT "S(12)=";P12*Q21/MD;"exp(j";V12+W12-VD;")"
1276 PRINT "S(21)=";P21*Q21/MD;"exp(j";V21+W21-VD;")"
1278 LET R=Q22* COS(W22)+Q12*Q21*P22/MD* COS (W12+W21+V22-VD)
1280 LET X=Q22* SIN(W22)+Q12*Q21*P22/MD* SIN (W12+W21+V22-VD)
1282 GOSUB 1288
1284 PRINT "S(22)=";M;"exp(j";V;")"
1286 GOTO 1314
1288 REM Subroutine 1288
1290 LET M= SQR (R*R+X*X)
1292 IF M=0 THEN  LET V=0: RETURN
1294 IF (X/M) >= .999999 THEN  LET V= PI /2: RETURN
1296 IF (X/M) <= -.999999 THEN  LET V=- PI /2: RETURN
1298 IF  ABS (R) <= 1E-09 THEN  GOTO 1306
1300 IF R<0 THEN  GOTO 1304
1302 LET V= ATN (X/R): RETURN
1304 LET V= PI + ATN (X/R): RETURN
1306 IF X=0 AND R>0 THEN  LET V=0: RETURN
1308 IF X=0 AND R<0 THEN  LET V= PI : RETURN
1310 IF X>0 THEN  LET V= PI /2- ATN (R/X): RETURN
1312 LET V=- PI /2+ ATN (R/X): RETURN
1314 REM Chain subroutine
1316 PRINT
1318 INPUT"Enter 'CONT':",N$
1320 IF N$="CONT" THEN CLS:CHAIN "MENU",10
1322 GOTO 1316
```

4.6 THE SMITH CHART TRANSFORMATION

The transmission line equation, known in the literature as the Smith chart transformation, is discussed in this section with regard to an electrical circuit, such as shown in Figure 4.7. The driving-point impedance Z_{in} at the plane 1-1′ is related to the terminating impedance Z_l by the equation:

$$Z_{in} = Z_0 \frac{Z_l + Z_0 \tanh(\gamma l)}{Z_0 + Z_l \tanh(\gamma l)} \tag{4.17}$$

where

Z_0 = the characteristic impedance of the transmission line of length l,
$\gamma = \alpha + j\beta$ = the propagation constant,
α = the attenuation constant, and
β = the phase constant,

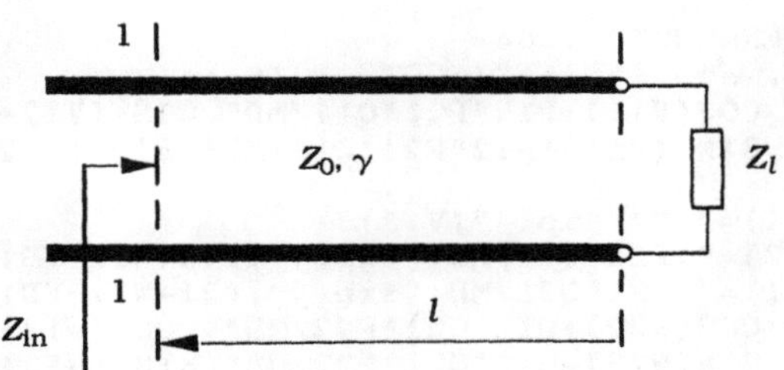

Figure 4.7 Transformation of the terminating impedance via the lossy TEM transmission line.

This equation derives its name from the fact that it is the basis for the Smith chart (see, e.g., [1,2,4,8]). Equation (4.17), after a simple rearrangement of the terms, assumes the form:

$$Z_{in} = Z_0 \frac{z_l + \tanh(\alpha l + j\beta l)}{1 + z_l \tanh(\alpha l + j\beta l)} \tag{4.18}$$

where $z_l = Z_l/Z_0 = |z_l| \exp(j\upsilon_l)$ is a normalized terminating impedance.

To present (4.18) in extended form (more convenient for calculations), we can use the following identity [7]:

$$\tanh(\alpha l + j\beta l) = \frac{\sinh(2\alpha l) + j\sin(2\beta l)}{\cosh(2\alpha l) + \cos(2\beta l)} \tag{4.19}$$

By substituting (4.19) into (4.18) we get:

$$Z_{in} = Z_0 \frac{r + r_l + j(x + x_l)}{1 + rr_l - xx_l + j(r_l x + x_l r)} \tag{4.20}$$

where

$$\begin{aligned} r_l &= |z_l| \cos(\upsilon_l) \\ x_l &= |z_l| \sin(\upsilon_l) \\ r &= \sinh(2\alpha l)/D \\ x &= \sin(2\beta l)/D \\ D &= \cosh(2\alpha l) + \cos(2\beta l) \end{aligned}$$

Note that the above approach can be easily generalized for the multisection stepped transmission line, as shown in Figure 4.8. To solve this problem, the impedance transforming procedure given by (4.20) is repeated several times for each of the individual sections. This can be done by means of the computer program SMITH. The calculations are made for the following input data:

n = the number of sections ($i = 1, 2, \ldots, n \geqslant 1$),
Z_{0i} = the characteristic impedance of the ith section in Ω,
a_i = the attenuation coefficient for the ith section in dB/m,
b_i = the phase constant for the ith section in rad/m,
l_i = the geometrical length of the ith section in m,
Z_l = the terminating load impedance in Ω.

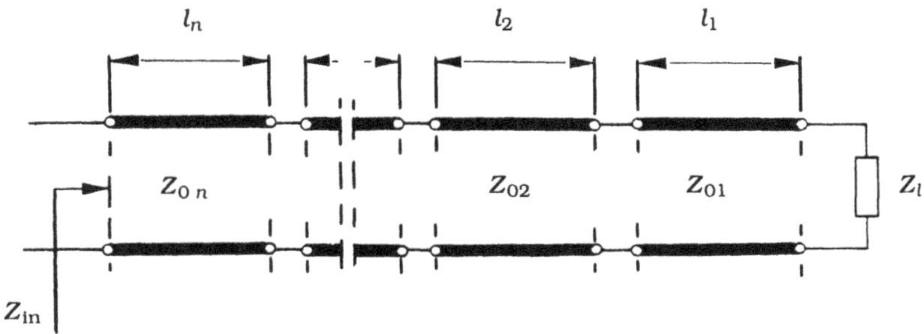

Figure 4.8 The multisection transmission line terminated by a complex impedance.

From the calculations, we obtain a value of the driving point impedance at the input plane 1-1′ (see Figure 4.8). The modulus and phase angle of this impedance are expressed in ohms and radians, respectively.

Check Calculations

The computer program SMITH can be verified by using the following data.

Input Data

$n = 2$
$Z_{01} = 50$, Ω
$Z_{02} = 60$, Ω
$a_1 = 0.3$, dB/m
$a_2 = 0.3$, dB/m
$b_1 = 1.2$, rad/m
$b_2 = 1.2$, rad/m
$l_1 = 1.0$, m
$l_2 = 0.6$, m
$Z_l = 75{,}000 \cdot \exp(j\,0.500)$, Ω

Results

$Z_{in} = 32.599 \cdot \exp(j\,0.108)$, Ω

Note: The attenuation coefficient, *a*, given in dB/m is related to the attenuation constant α by the following simple relation: $\alpha = a/(20 \log e) \approx 0.115129\, a$.

Listing of Program SMITH

```
1400 REM SMITH
1402 CLS
1404 PRINT "SMITH TRANSFORMATION"
1406 PRINT
1408 PRINT "Data:"
1410 PRINT
1412 INPUT "n=",N
1414 PRINT
1416 IF N<1 THEN  PRINT "Error:n<1": GOTO 1410
1418 IF  INT (N)=N THEN  GOTO 1424
1420 PRINT
1422 PRINT "Number n has to be an integer !": GOTO 1410
1424 DIM A(N)
1426 DIM B(N)
1428 DIM L(N)
1430 DIM Z(N)
1432 LET PI=3.141593
1434 FOR I=1 TO N
1436 PRINT USING"Zo(#)[ohm]=";I;: INPUT "",Z(I)
1438 PRINT USING"a(#)[dB/m]=";I;:INPUT "",A(I)
1440 PRINT USING"b(#)[rad/m]=";I;:INPUT "",B(I)
1442 PRINT USING"l(#)[m]=";I;:INPUT "",L(I)
1444 NEXT I
1446 FOR I=1 TO N
1448 IF Z(I) <= 0 THEN  PRINT "Error:Zo(";I;") <= 0": GOTO 1410
1450 IF A(I)<0 THEN  PRINT "Error:a(";I;")<0": GOTO 1410
1452 IF B(I)<0 THEN  PRINT "Error:b(";I;")<0": GOTO 1410
1454 IF L(I)<0 THEN  PRINT "Error:l(";I;")<0": GOTO 1410
1456 NEXT I
1458 PRINT
1460 INPUT ;"Zl[ohm]=",ML:INPUT ;" exp(j",VL:PRINT ")"
1462 IF ML<0 THEN  PRINT "Error:ABS(Zl)<0": GOTO 1460
1464 PRINT
1466 PRINT "Please wait !"
1468 LET MK=ML
1470 LET VK=VL
1472 FOR J=1 TO N
1474 LET MK=(MK+1E-12)/Z(J)
1476 LET D=.115129*A(J)
1478 LET SH=( EXP (2*D*L(J))- EXP (-2*D*L(J)))/2
1480 LET CH=( EXP (2*D*L(J))+ EXP (-2*D*L(J)))/2
1482 LET RT=SH/(CH+ COS (2*B(J)*L(J)))
1484 LET XT= SIN (2*B(J)*L(J))/(CH+ COS (2*B(J)*L(J)))
1486 LET R=RT+MK* COS(VK)
```

```
1488 LET X=XT+MK* SIN(VK)
1490 GOSUB 1518
1492 LET M=M1
1494 LET V=V1
1496 LET R=1+RT*MK* COS(VK)-XT*MK* SIN(VK)
1498 LET X=RT*MK* SIN(VK)+MK*XT* COS(VK)
1500 GOSUB 1518
1502 LET MK=Z(J)*M/M1
1504 LET VK=V-V1
1506 NEXT J
1508 PRINT
1510 PRINT "Results, see Fig. 4.8"
1512 PRINT
1514 PRINT "Zin[ohm]=";MK;" exp(j";VK;")"
1516 GOTO 1544
1518 REM Subroutine 1518
1520 LET M1= SQR (R*R+X*X)
1522 IF M1=0 THEN  LET V1=0: RETURN
1524 IF (X/M1) >= .999999 THEN  LET V1= PI /2: RETURN
1526 IF (X/M1) <= -.999999 THEN  LET V1=- PI /2: RETURN
1528 IF  ABS (R)<1E-09 THEN  GOTO 1536
1530 IF R<0 THEN  GOTO 1534
1532 LET V1= ATN (X/R): RETURN
1534 LET V1= PI + ATN (X/R): RETURN
1536 IF X=0 AND R>0 THEN  LET V1=0: RETURN
1538 IF X=0 AND R<0 THEN  LET V1= PI : RETURN
1540 IF X>0 THEN  LET V1= PI /2- ATN (R/X): RETURN
1542 LET V1=- PI /2+ ATN (R/X): RETURN
1544 REM Chain subroutine
1546 PRINT
1548 INPUT"Enter 'CONT':",N$
1550 IF N$="CONT" THEN CLS:CHAIN "MENU",10
1552 GOTO 1546
```

4.7 DESIGN OF THE STEPPED TRANSMISSION LINE BY THE CHAIN-MATRIX METHOD

Many microwave networks (e.g., impedance transformers and bandstop filters), are realized in the form of a multisection stepped transmission line. So far, the synthesis of these networks has been based primarily on the following insertion loss function [4,8,9]:

$$L(\theta) = 1 + h^2 T_n^2(x) \tag{4.21}$$

where

$$x = \begin{cases} \cos(\theta)/t, & \text{for quarter-wavelength transformers} \\ \sin(\theta)/t, & \text{for bandstop filters} \end{cases}$$

θ is the electrical length of the individual sections, h is the coefficient determining the maximum attenuation in the passband, t is the frequency coefficient, and $T_n(x)$ denotes the Chebyschev polynomial of the first kind of degree n.

The literature [10–13] has proved that by using the following insertion loss function:

$$L(\theta) = |T_{22}|^2 = 1 + h^2 T_n^2[(\cos\theta + m)/t] \tag{4.22}$$

to design the stepped impedance transformers and filters is possible with the most appropriate structures that more precisely satisfy the electrical and structural requirements. Hence, (4.22) is used here as a more general insertion-loss function of the lossless transmission line, the electrical scheme of which is shown in Figure 4.9.

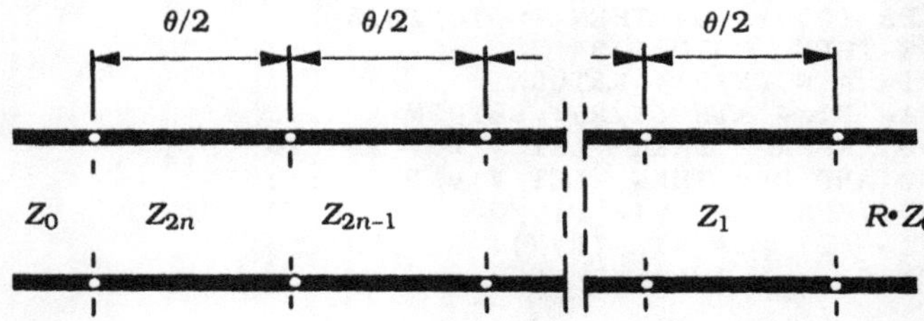

Figure 4.9 The multisection transmission line described by the insertion loss function (4.22).

The literature [10] has shown how the coefficients h, n, m, and t are related to the technical parameters of the impedance transformer. Similarly, in [12 and 13], these coefficients are expressed in terms of technical parameters of the stepped bandstop filters. If the values of the above mentioned coefficients are known, then it is possible to calculate the characteristic impedances of the individual sections of the line being considered. This can be done by means of the chain-matrix method presented below. The first step in this method consists in evaluating the polynomial $T_{22}(\theta)$ from (4.22), which satisfies the conditions for physical realization [4,8,14]. In the next step, the normalized *ABCD*-matrix, denoted as $[a]$, is derived on the basis of this polynomial. Finally, the matrix $[a]$ obtained in such a way allows us to calculate the characteristic impedances of the individual line sections.

Now, let us express (4.22) in the form of the following polynomial:

$$|T_{22}(q)|^2 = k_0 + k_2 q^2 + k_4 q^4 + \cdots + k_{4n} q^{4n} \tag{4.23}$$

where k_i are real coefficients for $i = 0, 2, 4, \ldots, 4n$ and the variable $q = \cos(\theta/2)$. Function (4.22) can be expressed in the form of (4.23) by calculating the discrete values of the variable $x_i = x(q_i)$ for which

$$1 + h^2 T_n^2(x_i) = 0 \tag{4.24}$$

According to [8], the roots x_i are:

$$x_i = \begin{cases} a_i + jb_i, & \text{for } i = 1 \text{ to } n \\ a_{2n+1-i} - jb_{2n+1-i}, & \text{for } i = n + 1 \text{ to } 2n \end{cases}$$

where:

$$a_i = \cosh(g)\cos[\pi(4i + 1)/(2n)],$$

$$b_i = \sinh(g)\sin[\pi(4i + 1)/(2n)], \text{ and}$$

$$g = (1/n)\ln\left[1/h + \sqrt{1 + (1/h)^2}\right]$$

The values of $x_i = x(q_i)$, satisfying equation (4.24), can be represented respectively by the $4n$ roots of the variable q calculated from:

$$q_i = \begin{cases} \sqrt{(x_i t - m + 1)/2}, & \text{for } i = 1 \text{ to } 2n \\ -q_{4n+1-i}, & \text{for } i = 2n + 1 \text{ to } 4n \end{cases}$$

To obtain the polynomial $T_{22}(q)$, satisfying conditions for realization, $2n$ roots q_i for which $\theta_i/2$ arguments lie in the upper half-plane of the variable θ have to be selected from the set of the roots q_i. This is achieved by introducing an additional variable p defined as [8]:

$$p = -j\cot(\theta/2)$$

The roots q_i are related to p_i as follows:

$$p_i = q_i/\sqrt{q_i^2 - 1} \tag{4.25}$$

Finally, the $2n$ roots p_i satisfying the condition $Re(p_i) < 0$ should be selected from the set of $4n$ values of the variable p. These roots allow us to write the desired polynominal $T_{22}(p)$ as

$$T_{22}(p) = \frac{\sqrt{L_m}}{(p^2 - 1)^n}(p - p_1)(p - p_2) \cdot \ldots \cdot (p - p_{2n}) \tag{4.26}$$

where $L_m = 1 + h^2T_n^2[(1 + m)/t]$. By using (4.25), the polynomial (4.26) can be expressed in terms of the variables $q = \cos(\theta/2)$ or $s = \sin(\theta/2)$. This polynomial can then be used to calculate the matrix $[a]$ of the designed line:

$$\begin{bmatrix} [Re(T_{22}) + T_{12}]/\sqrt{R}, & j\sqrt{R}\, Im(T_{22}) \\ jIm(T_{22})/\sqrt{R}, & \sqrt{R}[Re(T_{22}) - T_{12}] \end{bmatrix}$$

where R is the ratio of the terminating impedances (see Figure 4.9). The element T_{12} of the transfer scattering matrix $[T]$ results from the condition:

$$T_{12} = \sqrt{|T_{22}|^2 - 1}$$

because the designed stepped line is assumed to be a lossless network. Therefore, after a simple rearrangement, from (4.22) we obtain $T_{12} = hT_n(x)$.

To calculate the impedance of the final line section (see Figure 4.9), we multiply the matrix:

$$[a_{2n}]^{-1} = \begin{bmatrix} q, & -js\, z_{2n} \\ -js/z_{2n}, & q \end{bmatrix}$$

by the matrix $[a]$ of the entire line. Matrix $[a_{2n}]^{-1}$ is the inverse of the transmission line section with normalized characteristic impedance z_{2n} and the electrical length equal to $\theta/2$. As a result of the multiplication, the matrix $[a]$ of the line shortened by one section should be evaluated. This means that the degree of the polynomials contained in the elements a_{11} and a_{21} must decrease by not more than one. From this condition, we obtain an equation that has the impedance z_{2n} as its solution. By substituting the value of z_{2n} calculated in this way into expressions a_{11}, a_{12}, a_{21}, and a_{22}, we obtain the chain matrix $[a]$ of the line shortened by one section. By repeating the above procedure $2n$ times, we can evaluate the respective impedance values of all sections of the line being designed. The explicit design formulas obtained according to this approach are presented below.

Let us therefore assume that the values of coefficients h, m, R, t and Z_0 for the twelve-section line ($n = 6$) are known. In this case, twelve roots p_i (see (4.26)) can be presented as six pairs of conjugate complex numbers:

$$p_1 = p_2^* = r_a + jx_a = m_a \exp(j\phi_a)$$

$$p_3 = p_4^* = r_b + jx_b = m_b \exp(j\phi_b)$$

$$
\begin{aligned}
p_5 &= p_6^* = r_c + jx_c = m_c \exp(j\phi_c) \\
p_7 &= p_8^* = r_d + jx_d = m_d \exp(j\phi_d) \\
p_9 &= p_{10}^* = r_e + jx_e = m_e \exp(j\phi_e) \\
p_{11} &= p_{12}^* = r_f + jx_f = m_f \exp(j\phi_f)
\end{aligned}
$$

where

$$r_a < 0,\ r_b < 0,\ r_c < 0,\ r_d < 0,\ r_e < 0 \text{ and } r_f < 0.$$

Here, the asterisk * denotes the complex conjugate understood as follows:

$$p^* = a - jb \text{ if } p = a + jb$$

where $j = \sqrt{-1}$.

The real and imaginary components of these roots, together with h, m, R, t, and Z_0, make it possible to calculate the characteristic impedances Z_i, where $i = 1, 2, \ldots, 12$, of the individual sections. These calculations can be effectively performed by using the following formulas.

$$
\begin{aligned}
x_6 &= (1 + m)/t \\
c &= 1/\sqrt{R} \\
d &= m - 1 \\
l_m &= \sqrt{1 + h^2(32x_6^6 - 48x_6^4 + 18x_6^2 - 1)^2}
\end{aligned}
\tag{4.27}
$$

$$
\begin{aligned}
a(2,2) &= l_m(1 + m_a^2) \\
a(1,2) &= -l_m 2r_a \\
a(0,2) &= -l_m m_a^2
\end{aligned}
\tag{4.28}
$$

$$
\begin{aligned}
a(4,4) &= a(2,2)(1 + m_b^2) - 2r_b a(1,2) \\
a(3,4) &= a(1,2)(1 + m_b^2) - 2r_b a(2,2) \\
a(2,4) &= a(0,2)(1 + m_b^2) + 2r_b a(1,2) - m_b^2 a(2,2) \\
a(1,4) &= -2r_b a(0,2) - m_b^2 a(1,2) \\
a(0,4) &= -m_b^2 a(0,2)
\end{aligned}
\tag{4.29}
$$

$$
\begin{aligned}
a(6,6) &= a(4,4)(1 + m_c^2) - 2r_c a(3,4) \\
a(5,6) &= a(3,4)(1 + m_c^2) - 2r_c a(4,4)
\end{aligned}
$$

$$
\begin{aligned}
a(4,6) &= a(2,4)(1 + m_c^2) + 2r_c a(3,4) - 2r_c a(1,4) \\
&\quad - 2m_c^2 a(4,4) \\
a(3,6) &= a(1,4)(1 + m_c^2) - 2r_c a(2,4) - m_c^2 a(3,4) \\
a(2,6) &= a(0,4)(1 + m_c^2) + 2r_c a(1,4) - m_c^2 a(2,4) \\
a(1,6) &= -\ 2r_c a(0.4) - m_c^2 a(1,4) \\
a(0,6) &= -\ m_c^2 a(0,4)
\end{aligned}
\tag{4.30}
$$

$$
\begin{aligned}
a(8,8) &= a(6,6)(1 + m_d^2) - 2r_d a(5,6) \\
a(7,8) &= a(5,6)(1 + m_d^2) - 2r_d a(6,6) \\
a(6,8) &= a(4,6)(1 + m_d^2) + 2r_d a(5,6) - 2r_d a(3,6) \\
&\quad - m_d^2 a(6,6) \\
a(5,8) &= a(3,6)(1 + m_d^2) - 2r_d a(4,6) - m_d^2 a(5,6) \\
a(4,8) &= a(2,6)(1 + m_d^2) + 2r_d a(3.6) - 2r_d a(1,6) \\
&\quad - m_d^2 a(4,6) \\
a(3,8) &= a(1,6)(1 + m_d^2) - 2r_d a(2,6) - m_d^2 a(3,6) \\
a(2,8) &= a(0,6)(1 + m_d^2) + 2r_d a(1,6) - m_d^2 a(2,6) \\
a(1,8) &= -2r_d a(0,6) - m_d^2 a(1,6) \\
a(0.8) &= -\ m_d^2 a(0,6)
\end{aligned}
\tag{4.31}
$$

$$
\begin{aligned}
a(10,10) &= a(8,8)(1 + m_e^2) - 2r_e a(7,8) \\
a(9,10) &= a(7,8)(1 + m_e^2) - 2r_e a(8,8) \\
a(8,10) &= a(6,8)(1 + m_e^2) + 2r_e a(7,8) - 2r_e a(5,8) \\
&\quad - m_e^2 a(8,8) \\
a(7,10) &= a(5,8)(1 + m_e^2) - 2r_e a(6,8) - m_e^2 a(7,8) \\
a(6,10) &= a(4,8)(1 + m_e^2) + 2r_e a(5,8) - 2r_e a(3,8) \\
&\quad - m_e^2 a(6,8) \\
a(5,10) &= a(3,8)(1 + m_e^2) - 2r_e a(4,8) - m_e^2 a(5,8) \\
a(4,10) &= a(2,8)(1 + m_e^2) + 2r_e a(3,8) - 2r_e a(1,8) \\
&\quad - m_e^2 a(4,8) \\
a(3,10) &= a(1,8)(1 + m_e^2) - 2r_e a(2,8) - m_e^2 a(3,8) \\
a(2,10) &= a(0,8)(1 + m_e^2) + 2r_e a(1,8) - m_e^2 a(2,8)
\end{aligned}
\tag{4.32}
$$

$$a(1,10) = -2r_e a(0,8) - m_e^2 a(1,8)$$
$$a(0,10) = -m_e^2 a(0.8)$$

$$\begin{aligned}
a(12,12) &= a(10,10)(1 + m_f^2) - 2r_f a(9,10) \\
a(11,12) &= a(9,10)(1 + m_f^2) - 2r_f a(10,10) \\
a(10,12) &= a(8,10)(1 + m_f^2) + 2r_f a(9,10) - 2r_f a(7,10) \\
&\quad - m_f^2 a(10,10) \\
a(9,12) &= a(7,10)(1 + m_f^2) - 2r_f a(8,10) - m_f^2 a(9,10) \\
a(8,12) &= a(6,10)(1 + m_f^2) + 2r_f a(7,10) - 2r_f a(5,10) \\
&\quad - m_f^2 a(8,10)
\end{aligned} \tag{4.33}$$

$$\begin{aligned}
a(7,12) &= a(5,10)(1 + m_f^2) - 2r_f a(6,10) - m_f^2 a(7,10) \\
a(6,12) &= a(4,10)(1 + m_f^2) + 2r_f a(5,10) - 2r_f a(3,10) \\
&\quad - m_f^2 a(6,10) \\
a(5,12) &= a(3,10)(1 + m_f^2) - 2r_f a(4,10) - m_f^2 a(5,10) \\
a(4,12) &= a(2,10)(1 + m_f^2) + 2r_f a(3,10) - 2r_f a(1,10) \\
&\quad - m_f^2 a(4,10) \\
a(3,12) &= a(1,10)(1 + m_f^2) - 2r_f a(2,10) - m_f^2 a(3,10) \\
a(2,12) &= a(0,10)(1 + m_f^2) + 2r_f a(1,10) - m_f^2 a(2,10) \\
a(1,12) &= -2r_f a(0,10) - m_f^2 a(1,10)
\end{aligned}$$

$$a(0,12) = -\ m_f^2 a(0,10)$$

$$\begin{aligned}
w(12,12) &= c[a(12,12) + 2048h/t^6] \\
w(11,12) &= c \cdot a(11,12) \\
w(10,12) &= c[a(10,12) + 6144h\,d/t^6] \\
w(9,12) &= c \cdot a(9,12) \\
w(8,12) &= c[a(8,12) + 768h(10d^2 - t^2)/t^6] \\
w(7,12) &= c \cdot a(7,12) \\
w(6,12) &= c[a(6,12) + 512h(10d^3 - 3dt^2)/t^6] \\
w(5,12) &= c \cdot a(5,12) \\
w(4,12) &= c[a(4,12) + 384h(5d^4 - 3d^2t^2)/t^6 + 72h/t^2] \\
w(3,12) &= c \cdot a(3,12) \\
w(2,12) &= c[a(2,12) + 384h(d^5 - d^3t^2)/t^6 + 72hd/t^2]
\end{aligned} \tag{4.34}$$

$$w(1,12) = c \cdot a(1,12)$$
$$w(0,12) = c[a(0,12) + 16h(2d^6 - 3d^4t^2)/t^6 + 18hd^2/t^2 - h]$$

$$z_{12} = w(12,12)/w(11,12) \tag{4.35}$$

$$\begin{aligned}
w(11,11) &= w(10,12) + z_{12}w(11,12) - z_{12}w(9,12)\\
w(10,11) &= w(9,12) - w(10,12)/z_{12}\\
w(9,11) &= w(8,12) + z_{12}w(9,12) - z_{12}w(7,12)\\
w(8,11) &= w(7,12) - w(8,12)/z_{12}\\
w(7,11) &= w(6,12) + z_{12}w(7,12) - z_{12}w(5,12)\\
\\
w(6,11) &= w(5,12) - w(6,12)/z_{12}\\
w(5,11) &= w(4,12) + z_{12}w(5,12) - z_{12}w(3,12)\\
w(4,11) &= w(3,12) - w(4,12)/z_{12}\\
w(3,11) &= w(2,12) + z_{12}w(3,12) - z_{12}w(1,12)\\
w(2,11) &= w(1,12) - w(2,12)/z_{12}\\
w(1,11) &= w(0,12) + z_{12}w(1,12)\\
w(0,11) &= -w(0,12)/z_{12}
\end{aligned} \tag{4.36}$$

$$z_{11} = w(11,11)/w(10,11) \tag{4.37}$$

$$\begin{aligned}
w(10,10) &= w(9,11) + z_{11}w(10,11) - z_{11}w(8,11)\\
w(9,10) &= w(8,11) - w(9,11)/z_{11}\\
w(8,10) &= w(7,11) + z_{11}w(8,11) - z_{11}w(6,11)\\
w(7,10) &= w(6,11) - w(7,11)/z_{11}\\
w(6,10) &= w(5,11) + z_{11}w(6,11) - z_{11}w(4,11)\\
w(5,10) &= w(4,11) - w(5,11)/z_{11}\\
w(4,10) &= w(3,11) + z_{11}w(4,11) - z_{11}w(2,11)\\
w(3,10) &= w(2,11) - w(3,11)/z_{11}\\
w(2,10) &= w(1,11) + z_{11}w(2,11) - z_{11}w(0,11)\\
\\
w(1,10) &= w(0,11) - w(1,11)/z_{11}\\
w(0,10) &= z_{11}w(0,11)
\end{aligned} \tag{4.38}$$

$$z_{10} = w(10,10)/w(9,10) \tag{4.39}$$

$$\begin{aligned}
w(9,9) &= w(8,10) + z_{10}w(9,10) - z_{10}w(7,10) \\
w(8,9) &= w(7,10) - w(8,10)/z_{10} \\
w(7,9) &= w(6,10) + z_{10}w(7,10) - z_{10}w(5,10) \\
w(6,9) &= w(5,10) - w(6,10)/z_{10} \\
w(5,9) &= w(4,10) + z_{10}w(5,10) - z_{10}w(3,10) \\
w(4,9) &= w(3,10) - w(4,10)/z_{10} \\
w(3,9) &= w(2,10) + z_{10}w(3,10) - z_{10}w(1,10) \\
w(2,9) &= w(1,10) - w(2,10)/z_{10} \\
w(1,9) &= w(0,10) + z_{10}w(1,10) \\
w(0,9) &= -\ w(0,10)/z_{10}
\end{aligned} \tag{4.40}$$

$$z_9 = w(9,9)/w(8,9) \tag{4.41}$$

$$\begin{aligned}
w(8,8) &= w(7,9) + z_9w(8,9) - z_9w(6,9) \\
w(7,8) &= w(6,9) - w(7,9)/z_9 \\
w(6,8) &= w(5,9) + z_9w(6,9) - z_9w(4,9) \\
w(5,8) &= w(4,9) - w(5,9)/z_9 \\
w(4,8) &= w(3,9) + z_9w(4,9) - z_9w(2,9) \\
w(3,8) &= w(2,9) - w(3,9)/z_9 \\
w(2,8) &= w(1,9) + z_9w(2,9) - z_9w(0,9) \\
w(1,8) &= w(0,9) - w(1,9)/z_9 \\
w(0,8) &= z_9w(0,9)
\end{aligned} \tag{4.42}$$

$$z_8 = w(8,8)/w(7,8) \tag{4.43}$$

$$\begin{aligned}
w(7,7) &= w(6,8) + z_8w(7,8) - z_8w(5,8) \\
w(6,7) &= w(5,8) - w(6,8)/z_8 \\
w(5,7) &= w(4,8) + z_8w(5,8) - z_8w(3,8) \\
w(4,7) &= w(3,8) - w(4,8)/z_8 \\
w(3,7) &= w(2,8) + z_8w(3,8) - z_8w(1,8)
\end{aligned} \tag{4.44}$$

$$w(2,7) = w(1,8) - w(2,8)/z_8$$
$$w(1,7) = w(0,8) + z_8 w(1,8)$$
$$w(0,7) = - w(0,8)/z_8$$

$$z_7 = w(7,7)/w(6,7) \tag{4.45}$$

$$\begin{aligned}
w(6,6) &= w(5,7) + z_7 w(6,7) - z_7 w(4,7) \\
w(5,6) &= w(4,7) - w(5,7)/z_7 \\
w(4,6) &= w(3,7) + z_7 w(4,7) - z_7 w(2,7) \\
w(3,6) &= w(2,7) - w(3,7)/z_7 \\
w(2,6) &= w(1,7) + z_7 w(2,7) - z_7 w(0,7) \\
w(1,6) &= w(0,7) - w(1,7)/z_7 \\
w(0,6) &= z_7 w(0.7)
\end{aligned} \tag{4.46}$$

$$z_6 = w(6,6)/w(5,6) \tag{4.47}$$

$$\begin{aligned}
w(5,5) &= w(4,6) + z_6 w(5,6) - z_6 w(3,6) \\
w(4,5) &= w(3,6) - w(4,6)/z_6 \\
w(3,5) &= w(2,6) + z_6 w(3,6) - z_6 w(1,6)
\end{aligned} \tag{4.48}$$

$$w(2,5) = w(1,6) - w(2,6)/z_6$$
$$w(1,5) = w(0,6) + z_6 w(1,6)$$
$$w(0,5) = -w(0,6)/z_6$$

$$z_5 = w(5,5)/w(4,5) \tag{4.49}$$

$$\begin{aligned}
w(4,4) &= w(3,5) + z_5 w(4,5) - z_5 w(2,5) \\
w(3,4) &= w(2,5) - w(3,5)/z_5 \\
w(2,4) &= w(1,5) + z_5 w(2,5) - z_5 w(0,5) \\
w(1,4) &= w(0,5) - w(1,5)/z_5 \\
w(0,4) &= z_5 w(0,5)
\end{aligned} \tag{4.50}$$

$$z_4 = w(4,4)/w(3,4) \tag{4.51}$$

$$
\begin{aligned}
w(3,3) &= w(2,4) + z_4 w(3,4) - z_4 w(1,4) \\
w(2,3) &= w(1,4) - w(2,4)/z_4 \\
w(1,3) &= w(0,4) + z_4 w(1,4) \\
w(0,3) &= -w(0,4)/z_4
\end{aligned} \tag{4.52}
$$

$$z_3 = w(3,3)/w(2,3) \tag{4.53}$$

$$
\begin{aligned}
w(2,2) &= w(1,3) + z_3 w(2,3) - z_3 w(0,3) \\
w(1,2) &= w(0,3) - w(1,3)/z_3 \\
w(0,2) &= z_3 w(0,3)
\end{aligned} \tag{4.54}
$$

$$z_2 = w(2,2)/w(1,2) \tag{4.55}$$

$$
\begin{aligned}
w(1,1) &= w(0,2) + z_2 w(1,2) \\
w(0,1) &= -w(0,2)/z_2
\end{aligned} \tag{4.56}
$$

$$z_1 = w(1,1)/w(0,1) \tag{4.57}$$

Finally, the impedances $Z_1, Z_2, \ldots, Z_{2n}$ are:

$$Z_i = z_i \cdot Z_0 \text{ for } i = 1 \text{ to } 2n \tag{4.58}$$

Formulas (4.28) to (4.58) have been arranged so that they can be also used for designing two-, four-, six-, eight- and ten-section lines.

At $n = 5$, the line under consideration consists of ten sections. In this case, (4.26) includes ten roots p_i, which can be presented as five pairs of conjugate complex numbers:

$$
\begin{aligned}
p_1 &= p_2^* = r_a + jx_a = m_a \exp(j\phi_a) \\
p_3 &= p_4^* = r_b + jx_b = m_b \exp(j\phi_b) \\
p_5 &= p_6^* = r_c + jx_c = m_c \exp(j\phi_c) \\
p_7 &= p_8^* = r_d + jx_d = m_d \exp(j\phi_d) \\
p_9 &= p_{10}^* = r_e + jx_e = m_e \exp(j\phi_e)
\end{aligned}
$$

where $r_a < 0$, $r_b < 0$, $r_c < 0$, $r_d < 0$ and $r_e < 0$. The above given roots together with

$$x_5 = (1 + m)/t$$
$$c = 1/\sqrt{R}$$
$$d = m - 1$$

and

$$l_m = \sqrt{1 + h^2(16x_5^5 - 20x_5^3 + 5x_5)^2}$$

allow us to calculate the coefficients $w(i,10) = w'(i,10)$, where $i = 0, 1, 2, \ldots, 10$.

$$w'(10,10) = c[a(10,10) + 512h/t^5]$$
$$w'(9,10) = c \cdot a(9,10)$$
$$w'(8,10) = c[a(8,10) + 1280hd/t^5]$$
$$w'(7,10) = c \cdot a(7,10)$$
$$w'(6,10) = c[a(6,10) + 160h(8d^2 - t^2)/t^5]$$
$$w'(5,10) = c \cdot a(5,10)$$
$$w'(4,10) = c[a(4,10) + 80hd(8d^2 - 3t^2)/t^5]$$
$$w'(3,10) = c \cdot a(3,10)$$
$$w'(2,10) = c[a(2,10) + 40hd^2(4d^2 - 3t^2)/t^5 + 10h/t]$$
$$w'(1,10) = c \cdot a(1,10)$$
$$w'(0,10) = c[a(0,10) + 4hd^3(4d^2 - 5t^2)/t^5 + 5hd/t]$$

Here, we note that coefficients $a(i,10)$, where $i = 0$ to 10, are calculated by using (4.28) to (4.32). We can make further calculations according to (4.39) to (4.58).

For the eight-section line ($n = 4$), (4.26) includes eight roots p_i which can be presented as

$$p_1 = p_2^* = r_a + jx_a = m_a \exp(j\phi_a)$$
$$p_3 = p_4^* = r_b + jx_b = m_b \exp(j\phi_b)$$
$$p_5 = p_6^* = r_c + jx_c = m_c \exp(j\phi_c)$$
$$p_7 = p_8^* = r_d + jx_d = m_d \exp(j\phi_d)$$

where $r_a < 0$, $r_b < 0$, $r_c < 0$ and $r_d < 0$. These roots together with $x_4 = (1 + m)/t$, $c = 1/\sqrt{R}$, $d = m - 1$ and

$$l_m = \sqrt{1 + h^2(8x_4^4 - 8x_4^2 + 1)^2}$$

are used to calculate the coefficients $w(i,8) = w'(i,8)$, where $i = 0, 1, 2, \ldots, 8$.

$$\begin{aligned}
w'(8,8) &= c[a(8,8) + 128h/t^4] \\
w'(7,8) &= c \cdot a(7,8) \\
w'(6,8) &= c[a(6,8) + 256hd/t^4] \\
w'(5,8) &= c \cdot a(5,8) \\
w'(4,8) &= c[a(4,8) + 32h(6d^2 - t^2)t^4] \\
w'(3,8) &= c \cdot a(3,8) \\
w'(2,8) &= c[a(2,8) + 32hd\,(2d^2 - t^2)/t^4] \\
w'(1,8) &= c \cdot a(1,8) \\
w'(0,8) &= c[a(0,8) + 8hd^2(d^2 - t^2)/t^4 + h]
\end{aligned}$$

We can calculate the coefficients $a(i,8)$ included in $w'(i,8)$, where $i = 0$ to 8, by using formulas (4.28) to (4.31). For further calculations, we use (4.43) to (4.58).

If $n = 3$, (4.26) describes the six section line. In this case the roots p_i are

$$\begin{aligned}
p_1 &= p_2^* = r_a + jx_a = m_a \exp(j\phi_a) \\
p_3 &= p_4^* = r_b + jx_b = m_b \exp(j\phi_b) \\
p_5 &= p_6^* = r_c + jx_c = m_c \exp(j\phi_c)
\end{aligned}$$

where $r_a < 0$, $r_b < 0$ and $r_c < 0$. Above given roots together with $x_3 = (1 + m)/t$, $c = 1/\sqrt{R}$, $d = m - 1$ and

$$l_m = \sqrt{1 + h^2(4x_3^3 - 3x_3)^2}$$

make it possible to calculate the coefficients $w(i,6) = w'(i,6)$, where $i = 0$ to 6 and

$$\begin{aligned}
w'(6,6) &= c[a(6,6) + 32h/t^3] \\
w'(5,6) &= c \cdot a(5,6) \\
w'(4,6) &= c[a(4,6) + 48hd/t^3]
\end{aligned}$$

$$w'(3,6) = c \cdot a(3,6)$$
$$w'(2,6) = c[a(2,6) + h(24d^2 - 6t^2)/t^3]$$
$$w'(1,6) = c \cdot a(1,6)$$
$$w'(0,6) = c[a(0,6) + h(4d^3 - 3dt^2)/t^3]$$

We can calculate the coefficients $a(i,6)$ for $i = 0$ to 6 by using (4.28) to (4.30). Next, we use (4.47) to (4.58) for the further calculations.

The four-section line ($n = 2$) is characterized by (4.26), including four roots p_i which can be written as

$$p_1 = p_2^* = r_a + jx_a = m_a \exp(j\phi_a)$$
$$p_3 = p_4^* = r_b + jx_b = m_b \exp(j\phi_b)$$

where $r_a < 0$ and $r_b < 0$. For this case $x_2 = (1 + m)/t$, $c = 1/\sqrt{R}$, $d = m - 1$ and

$$l_m = \sqrt{1 + h^2(2x_2^2 - 1)^2}$$

The coefficients $w(i,4) = w'(i,4)$, where $i = 0$ to 4, are:

$$w'(4,4) = c[a(4,4) + 8h/t^2]$$
$$w'(3,4) = ca(3,4)$$
$$w'(2,4) = c[a(2,4) + 8hd/t^2]$$
$$w'(1,4) = c \cdot a(1,4)$$
$$w'(0,4) = c[a(0,4) + 2hd^2/t^2 - h]$$

We can calculate the coefficients $a(i,4)$ included in $w'(i,4)$, where $i = 0$ to 4, by using (4.28) and (4.29), and we can make further calculations according to (4.51) to (4.58).

Function (4.26) for the two-section line ($n = 1$) is determined by two roots p_i, namely:

$$p_1 = p_2^* = r_a + jx_a = m_a \exp(j\phi_a)$$

where $r_a < 0$. In this case $x_1 = (1 + m)/t$, $c = 1/\sqrt{R}$, $d = m - 1$ and

$$l_m = \sqrt{1 + h^2x_1^2}$$

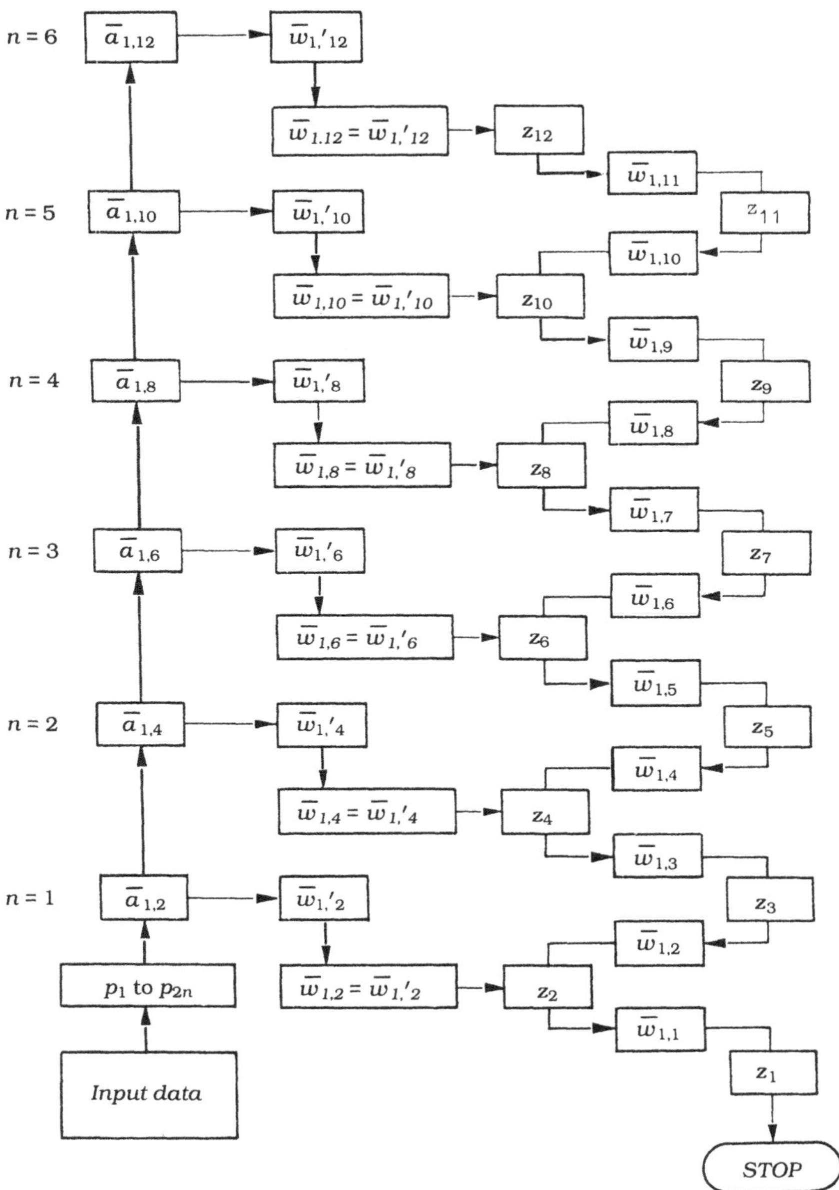

Figure 4.10 The schematic diagram of the calculation process described in Section 4.7.

The coefficients $w(i,2) = w'(i,2)$, where $i = 0$ to 2, are

$$w'(2,2) = c[a(2,2) + hd/t]$$
$$w'(1,2) = c \cdot a(1,2)$$
$$w'(0,2) = c[a(0,2) + 2h/t]$$

We can calculate the coefficients $a(i,2)$ included in $w'(i,2)$, where $i = 0$ to 2, by using (4.28), and make further calculations according to (4.55) to (4.58).

The calculation procedure presented above is outlined in Figure 4.10. We can make such calculations automatically by means of the computer program STL. For this purpose, we must introduce into the computer memory the values of $n \leq 6$, $R \geq 1$, $Z_0 > 0$, $h > 0$, $-1 < m \leq m_c$ [1] and $0 < t < 1$. If these values are suitable for synthesis (the line under synthesis is realizable), we obtain the characteristic impedances Z_i, where $i = 1, 2, \ldots, 2n$.

Check Calculations

The test data given in Table 4.2 can be used to verify the computer program STL.

Table 4.2
Some Illustrative Results Obtained with Program STL

	Z_0 = 25 Ω, R = 2 and h = 0.050062 (see Fig. 4.9)					
Parameters	*1*	*2*	*3*	*4*	*5*	*6*
n	2	3	4	4	5	6
m	0.335061	0	− 0.000044	0.061915	− 0.414307	0.045056
t	0.664944	0.707326	0.815402	0.865928	0.512427	0.951271
Z_1, Ω	44.704	43.488	44.704	45.767	35.526	46.748
Z_2, Ω	38.544	43.488	44.704	43.843	68.558	45.101
Z_3, Ω	32.430	35.354	38.543	39.853	22.868	43.081
Z_4, Ω	27.961	35.354	38.545	37.356	75.840	40.972
Z_5, Ω	—	28.743	32.429	33.461	18.315	38.674
Z_6, Ω	—	28.743	32.431	31.365	68.249	36.471
Z_7, Ω	—	—	27.961	28.510	16.482	34.273
Z_8, Ω	—	—	27.962	27.312	54.662	32.321
Z_9, Ω	—	—	—	—	18.233	30.508
Z_{10}, Ω	—	—	—	—	35.186	29.014
Z_{11}, Ω	—	—	—	—	—	27.715
Z_{12}, Ω	—	—	—	—	—	26.739

[1]The meaning of the parameter m_c is explained in Section 5.3.

Listing of Program STL

```
1600 REM STL
1602 CLS:DEFDBL A-Z:DEFINT I,J
1604 PRINT "STEPPED TRANSMISSION LINE"
1606 PRINT
1608 PRINT "Data:"
1610 PRINT
1612 INPUT "n=",N
1614 INPUT "Zo[ohm]=",ZO1
1616 INPUT "R=",RI
1618 INPUT "h=",H
1620 INPUT "m=",M
1622 INPUT "t=",T
1624 PRINT
1626 LET PI=3.141592653#
1628 IF ZO1 <= 0 THEN  PRINT "Error:Zo <= 0": GOTO 1610
1630 IF RI<1 THEN  PRINT "Assume R >= 1": GOTO 1610
1632 IF H <= 0 THEN  PRINT "Error:h <= 0": GOTO 1610
1634 IF T <= 0 THEN  PRINT "Error:t <= 0": GOTO 1610
1636 IF T >= 1 THEN  PRINT "Error:t >= 1": GOTO 1610
1638 IF M <= -.99 THEN  PRINT "m <= -.99": GOTO 1610
1640 LET LR=(RI+1)*(RI+1)/(4*RI)
1642 LET A= SQR (LR-1)/H
1644 LET X=(1+M)/T
1646 IF N=1 THEN  LET TN=X: GOTO 1660
1648 IF N=2 THEN  LET TN=2*X*X-1: GOTO 1660
1650 IF N=3 THEN  LET TN=4*X*X*X-3*X: GOTO 1660
1652 IF N=4 THEN  LET TN=8*X*X*X*X-8*X*X+1: GOTO 1660
1654 IF N=5 THEN  LET TN=16*X*X*X*X*X-20*X*X*X+5*X: GOTO 1660
1656 IF N=6 THEN  LET TN=32*X*X*X*X*X*X-48*X*X*X*X+18*X*X-1: GO
TO 1660
1658 PRINT "Number n takes only values 1, 2, 3, 4, 5 or 6": GOT
O 1610
1660 IF  ABS (A-TN) <= .001 THEN  GOTO 1664
1662 PRINT "The input data should satisfy the equation: Tn((1+m
)/t)-(R-1)/(2*h* SQR (R))=0": GOTO 1606
1664 PRINT "Please wait !"
1666 PRINT
1668 GOSUB 1688
1670 PRINT
1672 PRINT "Results, see Fig. 4.9"
1674 PRINT
1676 FOR I=1 TO 2*N
1678 PRINT "Z(";I;")[ohm]=";USING"###.######";ZO1*Z(I)
1680 NEXT I
1682 PRINT
1684 PRINT USING "O1[rad]=##.######"; ATN((T-M)/SQR(1-(T-M)*(T-
M)))
1686 GOTO 2366
1688 REM Procedure STL
1690 LET D=M-1
1692 LET C=1/ SQR(RI)
1694 DIM A(2*(N+1))
```

```
1696 DIM B(2*(N+1))
1698 DIM D(4*(N+1))
1700 DIM F(4*(N+1))
1702 DIM R(4*(N+1))
1704 DIM X(4*(N+1))
1706 DIM P(4*(N+1))
1708 DIM Z(2*(N+1))
1710 LET G= LOG(1/H+ SQR (1+(1/H)^2))/N
1712 LET CH=( EXP(G)+ EXP(-G))/2
1714 LET SH=( EXP(G)- EXP(-G))/2
1716 FOR I=1 TO N
1718 LET A(I)=CH* COS ( PI *(4*I+1)/(2*N))
1720 LET B(I)=SH* SIN ( PI *(4*I+1)/(2*N))
1722 NEXT I
1724 FOR I=N+1 TO 2*N
1726 LET A(I)=A(2*N+1-I)
1728 LET B(I)=-B(2*N+1-I)
1730 NEXT I
1732 FOR I=1 TO 2*N
1734 LET R=(A(I)*T-M+1)/2
1736 LET X=B(I)*T/2
1738 GOSUB 2340
1740 LET D(I)= SQR (M1)
1742 LET F(I)=F1/2
1744 NEXT I
1746 FOR I=2*N+1 TO 4*N
1748 LET D(I)=D(4*N+1-I)
1750 LET F(I)=F(4*N+1-I)+ PI
1752 NEXT I
1754 FOR I=1 TO 4*N
1756 LET R=D(I)*D(I)* COS (2*F(I))-1
1758 LET X=D(I)*D(I)* SIN (2*F(I))
1760 GOSUB 2340
1762 LET MP=D(I)/ SQR (M1)
1764 LET FP=F(I)-F1/2
1766 LET R(I)=MP* COS(FP)
1768 LET X(I)=MP* SIN(FP)
1770 NEXT I
1772 LET RA= ABS(R(1))
1774 LET XA= ABS(X(1))
1776 IF N=1 THEN  GOTO 1846
1778 FOR I=2 TO 4*N
1780 IF  ABS(R(I))<> RA THEN  LET RB= ABS(R(I)): LET XB= ABS(X(
I)): LET J=I+1: GOTO 1784
1782 NEXT I
1784 IF N=2 THEN  GOTO 1852
1786 FOR I=J TO 4*N
1788 IF  ABS(R(I))<> RA AND  ABS(R(I))<> RB THEN  LET RC= ABS(R
(I)): LET XC= ABS(X(I)): LET J=I+1: GOTO 1792
1790 NEXT I
1792 IF N=3 THEN  GOTO 1860
1794 FOR I=J TO 4*N
1796 IF  ABS(R(I))=RA THEN  NEXT I
1798 IF  ABS(R(I))=RB THEN  NEXT I
1800 IF  ABS(R(I))=RC THEN  NEXT I
```

```
1802 LET RD= ABS(R(I))
1804 LET XD= ABS(X(I))
1806 LET J=I+1
1808 IF N=4 THEN  GOTO 1870
1810 FOR I=J TO 4*N
1812 IF  ABS(R(I))=RA THEN  NEXT I
1814 IF  ABS(R(I))=RB THEN  NEXT I
1816 IF  ABS(R(I))=RC THEN  NEXT I
1818 IF  ABS(R(I))=RD THEN  NEXT I
1820 LET RE= ABS(R(I))
1822 LET XE= ABS(X(I))
1824 LET J=I+1
1826 IF N=5 THEN  GOTO 1882
1828 FOR I=J TO 4*N
1830 IF  ABS(R(I))=RA THEN  NEXT I
1832 IF  ABS(R(I))=RB THEN  NEXT I
1834 IF  ABS(R(I))=RC THEN  NEXT I
1836 IF  ABS(R(I))=RD THEN  NEXT I
1838 IF  ABS(R(I))=RE THEN  NEXT I
1840 LET RF= ABS(R(I))
1842 LET XF= ABS(X(I))
1844 GOTO 1896
1846 REM n=1
1848 LET RA=-RA
1850 GOTO 1910
1852 REM n=2
1854 LET RA=-RA
1856 LET RB=-RB
1858 GOTO 1910
1860 REM n=3
1862 LET RA=-RA
1864 LET RB=-RB
1866 LET RC=-RC
1868 GOTO 1910
1870 REM n=4
1872 LET RA=-RA
1874 LET RB=-RB
1876 LET RC=-RC
1878 LET RD=-RD
1880 GOTO 1910
1882 REM n=5
1884 LET RA=-RA
1886 LET RB=-RB
1888 LET RC=-RC
1890 LET RD=-RD
1892 LET RE=-RE
1894 GOTO 1910
1896 REM n=6
1898 LET RA=-RA
1900 LET RB=-RB
1902 LET RC=-RC
1904 LET RD=-RD
1906 LET RE=-RE
1908 LET RF=-RF
1910 LET SL= SQR (1+H*H*TN*TN)
```

```
1912 LET WA=RA*RA+XA*XA
1914 LET A22=SL*(1+WA)
1916 LET A12=-SL*2*RA
1918 LET A02=-SL*WA
1920 IF N>1 THEN  GOTO 1930
1922 LET W22=C*(A22+H*2/T)
1924 LET W12=C*A12
1926 LET W02=C*(A02+H*D/T)
1928 GOTO 2326
1930 LET WB=RB*RB+XB*XB
1932 LET A44=SL*((1+WA)*(1+WB)+4*RA*RB)
1934 LET A34=-2*SL*(RB*(1+WA)+RA*(1+WB))
1936 LET A24=-SL*(2*WA*WB+WA+WB+4*RA*RB)
1938 LET A14=2*SL*(RB*WA+RA*WB)
1940 LET A04=SL*WA*WB
1942 IF N>2 THEN  GOTO 1956
1944 LET W44=C*(A44+8*H/(T*T))
1946 LET W34=C*A34
1948 LET W24=C*(A24+8*H*D/(T*T))
1950 LET W14=C*A14
1952 LET W04=C*(A04+H*2*D*D/(T*T)-H)
1954 GOTO 2304
1956 LET WC=RC*RC+XC*XC
1958 LET A66=A44*(1+WC)-2*RC*A34
1960 LET A56=A34*(1+WC)-2*RC*A44
1962 LET A46=A24*(1+WC)+2*RC*A34-2*RC*A14-A44*WC
1964 LET A36=A14*(1+WC)-2*RC*A24-A34*WC
1966 LET A26=A04*(1+WC)+2*RC*A14-A24*WC
1968 LET A16=-(2*RC*A04+A14*WC)
1970 LET A06=-A04*WC
1972 IF N>3 THEN  GOTO 1990
1974 LET W66=C*(A66+32*H/(T*T*T))
1976 LET W56=C*A56
1978 LET W46=C*(A46+48*H*D/(T*T*T))
1980 LET W36=C*A36
1982 LET W26=C*(A26+24*H*D*D/(T*T*T)-6*H/T)
1984 LET W16=C*A16
1986 LET W06=C*(A06+4*H*D*D*D/(T*T*T)-3*H*D/T)
1988 GOTO 2274
1990 LET WD=RD*RD+XD*XD
1992 LET A88=A66*(1+WD)-2*RD*A56
1994 LET A78=A56*(1+WD)-2*RD*A66
1996 LET A68=A46*(1+WD)+2*RD*A56-2*RD*A36-WD*A66
1998 LET A58=A36*(1+WD)-2*RD*A46-WD*A56
2000 LET A48=A26*(1+WD)+2*RD*A36-2*RD*A16-WD*A46
2002 LET A38=A16*(1+WD)-2*RD*A26-WD*A36
2004 LET A28=A06*(1+WD)+2*RD*A16-WD*A26
2006 LET A18=-2*RD*A06-WD*A16
2008 LET A08=-WD*A06
2010 IF N>4 THEN  GOTO 2032
2012 LET W88=C*(A88+128*H/(T*T*T*T))
2014 LET W78=C*A78
2016 LET W68=C*(A68+256*H*D/(T*T*T*T))
2018 LET W58=C*A58
2020 LET W48=C*(A48+32*H*(6*D*D-T*T)/(T*T*T*T))
```

```
2022 LET W38=C*A38
2024 LET W28=C*(A28+32*H*D*(2*D*D-T*T)/(T*T*T*T))
2026 LET W18=C*A18
2028 LET W08=C*(A08+8*H*D*D*(D*D-T*T)/(T*T*T*T)+H)
2030 GOTO 2236
2032 LET WE=RE*RE+XE*XE
2034 LET A1010=A88*(1+WE)-2*RE*A78
2036 LET A910=A78*(1+WE)-2*RE*A88
2038 LET A810=A68*(1+WE)+2*RE*A78-2*RE*A58-WE*A88
2040 LET A710=A58*(1+WE)-2*RE*A68-WE*A78
2042 LET A610=A48*(1+WE)+2*RE*A58-2*RE*A38-WE*A68
2044 LET A510=A38*(1+WE)-2*RE*A48-WE*A58
2046 LET A410=A28*(1+WE)+2*RE*A38-2*RE*A18-WE*A48
2048 LET A310=A18*(1+WE)-2*RE*A28-WE*A38
2050 LET A210=A08*(1+WE)+2*RE*A18-WE*A28
2052 LET A110=-2*RE*A08-WE*A18
2054 LET A010=-WE*A08
2056 IF N>5 THEN  GOTO 2082
2058 LET W1010=C*(A1010+512*H/(T*T*T*T*T))
2060 LET W910=C*A910
2062 LET W810=C*(A810+1280*H*D/(T*T*T*T*T))
2064 LET W710=C*A710
2066 LET W610=C*(A610+160*H*(8*D*D-T*T)/(T*T*T*T*T))
2068 LET W510=C*A510
2070 LET W410=C*(A410+80*H*D*(8*D*D-3*T*T)/(T*T*T*T*T))
2072 LET W310=C*A310
2074 LET W210=C*(A210+40*H*D*D*(4*D*D-3*T*T)/(T*T*T*T*T)+10*H/T
)
2076 LET W110=C*A110
2078 LET W010=C*(A010+H*D*(16*D*D*D*D/(T*T*T*T*T)-20*D*D/(T*T*T
)+5/T))
2080 GOTO 2190
2082 LET WF=RF*RF+XF*XF
2084 LET A1212=A1010*(1+WF)-2*RF*A910
2086 LET A1112=A910*(1+WF)-2*RF*A1010
2088 LET A1012=A810*(1+WF)+2*RF*A910-2*RF*A710-WF*A1010
2090 LET A912=A710*(1+WF)-2*RF*A810-WF*A910
2092 LET A812=A610*(1+WF)+2*RF*A710-2*RF*A510-WF*A810
2094 LET A712=A510*(1+WF)-2*RF*A610-WF*A710
2096 LET A612=A410*(1+WF)+2*RF*A510-2*RF*A310-WF*A610
2098 LET A512=A310*(1+WF)-2*RF*A410-WF*A510
2100 LET A412=A210*(1+WF)+2*RF*A310-2*RF*A110-WF*A410
2102 LET A312=A110*(1+WF)-2*RF*A210-WF*A310
2104 LET A212=A010*(1+WF)+2*RF*A110-WF*A210
2106 LET A112=-2*RF*A010-WF*A110
2108 LET A012=-WF*A010
2110 LET W1212=C*(A1212+H*2048/(T*T*T*T*T*T))
2112 LET W1112=C*A1112
2114 LET W1012=C*(A1012+H*D*6144/(T*T*T*T*T*T))
2116 LET W912=C*A912
2118 LET W812=C*(A812+H*768*(10*D*D-T*T)/(T*T*T*T*T*T))
2120 LET W712=C*A712
2122 LET W612=C*(A612+H*512*(10*D*D*D-3*D*T*T)/(T*T*T*T*T*T))
2124 LET W512=C*A512
2126 LET W412=C*(A412+H*384*(5*D*D*D*D-3*D*D*T*T)/(T*T*T*T*T*T)
```

```
+72*H/(T*T))
2128 LET W312=C*A312
2130 LET W212=C*(A212+H*384*(D*D*D*D-D*D*D*T*T)/(T*T*T*T*T*T)
+72*H*D/(T*T))
2132 LET W112=C*A112
2134 LET W012=C*(A012+H*16*(2*D*D*D*D*D*D-3*D*D*D*D*T*T)/(T*T*T
*T*T*T)+18*H*D*D/(T*T)-H)
2136 LET Z12=W1212/W1112
2138 LET Z(12)=Z12
2140 LET W1111=W1012+Z12*W1112-Z12*W912
2142 LET W1011=W912-W1012/Z12
2144 LET W911=W812+Z12*W912-Z12*W712
2146 LET W811=W712-W812/Z12
2148 LET W711=W612+Z12*W712-Z12*W512
2150 LET W611=W512-W612/Z12
2152 LET W511=W412+Z12*W512-Z12*W312
2154 LET W411=W312-W412/Z12
2156 LET W311=W212+Z12*W312-Z12*W112
2158 LET W211=W112-W212/Z12
2160 LET W111=W012+Z12*W112
2162 LET W011=-W012/Z12
2164 LET Z11=W1111/W1011
2166 LET Z(11)=Z11
2168 LET W1010=W911+Z11*W1011-Z11*W811
2170 LET W910=W811-W911/Z11
2172 LET W810=W711+Z11*W811-Z11*W611
2174 LET W710=W611-W711/Z11
2176 LET W610=W511+Z11*W611-Z11*W411
2178 LET W510=W411-W511/Z11
2180 LET W410=W311+Z11*W411-Z11*W211
2182 LET W310=W211-W311/Z11
2184 LET W210=W111+Z11*W211-Z11*W011
2186 LET W110=W011-W111/Z11
2188 LET W010=Z11*W011
2190 LET Z10=W1010/W910
2192 LET Z(10)=Z10
2194 LET W99=W810+Z10*W910-Z10*W710
2196 LET W89=W710-W810/Z10
2198 LET W79=W610+Z10*W710-Z10*W510
2200 LET W69=W510-W610/Z10
2202 LET W59=W410+Z10*W510-Z10*W310
2204 LET W49=W310-W410/Z10
2206 LET W39=W210+Z10*W310-Z10*W110
2208 LET W29=W110-W210/Z10
2210 LET W19=W010+Z10*W110
2212 LET W09=-W010/Z10
2214 LET Z9=W99/W89
2216 LET Z(9)=Z9
2218 LET W88=W79+Z9*W89-Z9*W69
2220 LET W78=W69-W79/Z9
2222 LET W68=W59+Z9*W69-Z9*W49
2224 LET W58=W49-W59/Z9
2226 LET W48=W39+Z9*W49-Z9*W29
2228 LET W38=W29-W39/Z9
2230 LET W28=W19+Z9*W29-Z9*W09
2232 LET W18=W09-W19/Z9
```

```
2234 LET W08=Z9*W09
2236 LET Z8=W88/W78
2238 LET Z(8)=Z8
2240 LET W77=W68+Z8*W78-Z8*W58
2242 LET W67=W58-W68/Z8
2244 LET W57=W48+Z8*W58-Z8*W38
2246 LET W47=W38-W48/Z8
2248 LET W37=W28+Z8*W38-Z8*W18
2250 LET W27=W18-W28/Z8
2252 LET W17=W08+Z8*W18
2254 LET W07=-W08/Z8
2256 LET Z7=W77/W67
2258 LET Z(7)=Z7
2260 LET W66=W57+Z7*W67-Z7*W47
2262 LET W56=W47-W57/Z7
2264 LET W46=W37+Z7*W47-Z7*W27
2266 LET W36=W27-W37/Z7
2268 LET W26=W17+Z7*W27-Z7*W07
2270 LET W16=W07-W17/Z7
2272 LET W06=Z7*W07
2274 LET Z6=W66/W56
2276 LET Z(6)=Z6
2278 LET W55=W46+Z6*W56-Z6*W36
2280 LET W45=W36-W46/Z6
2282 LET W35=W26+Z6*W36-Z6*W16
2284 LET W25=W16-W26/Z6
2286 LET W15=W06+Z6*W16
2288 LET W05=-W06/Z6
2290 LET Z5=W55/W45
2292 LET Z(5)=Z5
2294 LET W44=W35+Z5*W45-Z5*W25
2296 LET W34=W25-W35/Z5
2298 LET W24=W15+Z5*W25-Z5*W05
2300 LET W14=W05-W15/Z5
2302 LET W04=Z5*W05
2304 LET Z4=W44/W34
2306 LET Z(4)=Z4
2308 LET W33=W24+Z4*W34-Z4*W14
2310 LET W23=W14-W24/Z4
2312 LET W13=W04+Z4*W14
2314 LET W03=-W04/Z4
2316 LET Z3=W33/W23
2318 LET Z(3)=Z3
2320 LET W22=W13+Z3*W23-Z3*W03
2322 LET W12=W03-W13/Z3
2324 LET W02=W03*Z3
2326 LET Z2=W22/W12
2328 LET Z(2)=Z2
2330 LET W11=W02+Z2*W12
2332 LET W01=-W02/Z2
2334 LET Z1=W11/W01
2336 LET Z(1)=Z1
2338 RETURN
2340 REM Subroutine 2340
2342 LET M1= SQR (R*R+X*X)
2344 IF M1=0 THEN  LET F1=0: RETURN
```

```
2346 IF (X/M1) >= .999999 THEN  LET F1= PI /2: RETURN
2348 IF (X/M1) <= -.999999 THEN  LET F1=- PI /2: RETURN
2350 IF  ABS (R)<1E-09 THEN  GOTO 2358
2352 IF R<0 THEN  GOTO 2356
2354 LET F1= ATN (X/R): RETURN
2356 LET F1= PI + ATN (X/R): RETURN
2358 IF X=0 AND R>0 THEN  LET F1=0: RETURN
2360 IF X=0 AND R<0 THEN  LET F1= PI : RETURN
2362 IF X>0 THEN  LET F1= PI /2- ATN (R/X): RETURN
2364 LET F1=- PI /2+ ATN (R/X): RETURN
2366 REM Chain subroutine
2368 PRINT
2370 INPUT"Enter 'CONT':",N$
2372 IF N$="CONT" THEN CLS:CHAIN "MENU",10
2374 GOTO 2368
```

REFERENCES

[1] Ware, L.A., and H.R. Reed, *Communication Circuits,* New York: John Wiley and Sons, 1949.

[2] Carlin, H.J., and A.B. Giordano, *Network Theory. An Introduction to Reciprocal and Nonreciprocal Circuits,* Englewood Cliffs, NJ: Prentice-Hall, 1964.

[3] *S-Parameter Techniques for Faster, More Accurate Network Design,* Hewlett-Packard, Application Note 95-1, September 1968.

[4] Matthaei, G.L., L. Young, and E.M.T. Jones, *Microwave Filters, Impedance Matching Networks and Coupling Structures,* Norwood, MA: Artech House, 1980.

[5] Gupta, K.C., R. Garg, and R. Chadha, *Computer-Aided Design of Microwave Circuits,* Norwood, MA: Artech House, 1981.

[6] Besser, L., and C. Hsieh, "Series and Parallel Addition of Two-ports via S-Parameters," *Microwave Journal,* Vol. 18, No. 4, April 1975, pp. 53–56.

[7] Dwight, H.B., *Tables of Integrals and Other Mathematical Data,* New York: The Macmillan, 1961.

[8] Feldstein, A.L., and L.R. Javich, *Synthesis of Microwave Two-Ports and Four-Ports* (in Russian), Moscow: Sviaz, 1971.

[9] Matthaei, G.L., "Short-Step Chebyschev Impedance Transformers," *IEEE Trans., Microwave Theory and Techniques,* Vol. MTT-14, No. 8, August 1966, pp. 372–383.

[10] Rosloniec, S., "A New Approach to the Synthesis of Microwave Impedance Transformers," *International Journal of Electronics,* Vol. 64, No. 6, June 1988, pp. 947–954.

[11] Rosloniec, S., "Design of the R-transformer on the Basis of the Insertion Loss Function," *International Journal of Electronics,* Vol. 64, No. 3, March 1988, pp. 385–394.

[12] Rosloniec, S., "Application of the R-transformer to the Design of Microwave Harmonic Filters," *Archiv für Elektronik und Übertragungstechnik,* AEÜ-41, Heft 4, Juli-August 1987, pp. 251–253.

[13] Rosloniec, S., "Synthesis of Stepped Impedance Transforming Band-stop Filters (in Russian)," *Journal of USSR Academy of Science,* Radio Engineering and Electronics, Vol. 34, No. 2, February 1989, pp. 305–310.

[14] Riblet, H.J., "General Synthesis of Quarter-Wave Impedance Transformers," *IRE Trans., Microwave Theory and Techniques,* Vol. MTT-5, No. 1, January 1957, pp. 36–43.

[15] White, J.F., "The Smith Chart: An Endangered Species?," *Microwave Journal,* Vol. 22, No. 11, November 1979, pp. 49–54.

[16] Chen, W.K., *Theory and Design of Broadband Matching Networks,* New York: Pergamon Press, 1976.

Chapter 5
CAD ALGORITHMS FOR SOME LINEAR MICROWAVE CIRCUITS

5.1 INTRODUCTION

This chapter provides the reader with some CAD algorithms intended for design of common use UHF and microwave circuits. The list of circuits under discussion includes such passive devices as broadband transmission line impedance transformers, branch and coupled transmission line directional couplers, three-port hybrid power dividers (summers), resistance attenuators, narrow-band impedance matching circuits, and different types of reactance filters. As a rule, these circuits have maximally flat or Chebyschev frequency responses. At microwave frequencies, they are usually made of lumped *RLC* elements and commensurate TEM transmission line segments. Therefore, the algorithms we present here allow us to calculate some geometrical dimensions of the circuits being designed, which is possible using the computer programs described in Chapter 6.

In general, these CAD algorithms are presented in the following way. The short description and electrical scheme of the circuit are given first. Next, we present the closed-form design formulas, together with the methods for solving them. In most cases, we present a design algorithm illustrated both by a flow diagram and some typical check calculations. An integral part of every design algorithm is the corresponding computer program written in BASIC. Usually, the listing of this program is given at the end of the section. The author's intention has been to present all of the above-mentioned CAD algorithms so that they are readily understood and clearly applicable.

5.2 QUARTER-WAVELENGTH IMPEDANCE TRANSFORMERS

The computer program Q-IT makes possible the design of quarter-wavelength impedance transformers with maximally flat and Chebyschev frequency responses. The electrical scheme and structural outline of these transformers are shown in

Figure 5.1. The name "quarter-wavelength" implies that all sections of the circuits under consideration have the same electrical length, equal to $\pi/2$ radians at the center frequency of an assumed matching band. The characteristic impedances of the individual sections can generally be calculated on the basis of Collin's equations or by using the Riblet's technique (see, e.g., [1] to [6]). Collin's equations, presented below, allow us to design quarter-wavelength transformers of up to four sections.

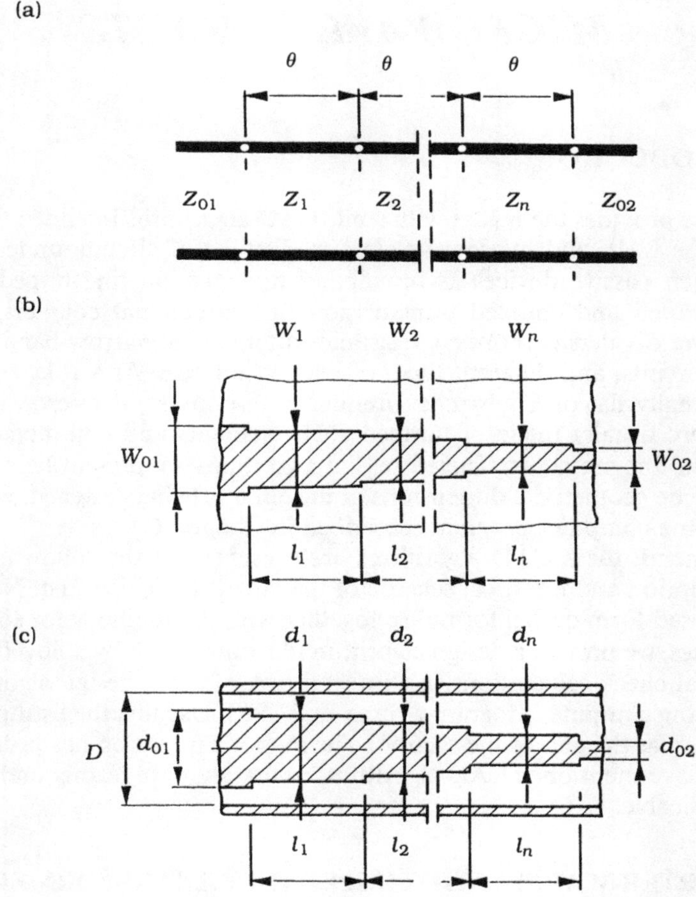

Figure 5.1 The stepped line impedance transformer: (a) electrical scheme, (b) microstrip line configuration and (c) coaxial line longitudinal section.

The characteristic impedance Z_k of kth section generally is

$$Z_k = Z_{k-1}V_k, \quad \text{for } k = 1, 2, 3, \text{and } 4 \tag{5.1}$$

where Z_{k-1} is the characteristic impedance of $(k-1)$th section. According to formula (5.1), Z_0 is a terminating resistance, which is to be matched to a resistance $R \cdot Z_0$, where $R \geqslant 1$. In Figure 5.1, these terminating resistances are denoted by Z_{01} and Z_{02}, respectively. The values of coefficients V_k are calculated on the basis of the following equations.

For maximally flat impedance transformers with one, two, three, and four sections, we have

(1)
$$\begin{aligned} n &= 1 \\ V_1 &= \sqrt{R} \end{aligned} \tag{5.2}$$

(2)
$$\begin{aligned} n &= 2 \\ V_1 &= \sqrt[4]{R} \\ V_2 &= \sqrt{R} \end{aligned} \tag{5.3}$$

(3)
$$\begin{aligned} n &= 3 \\ V_1^2 + 2V_1\sqrt{R} - 2\sqrt{R}/V_1 - R/V_1^2 &= 0 \\ V_2 &= \sqrt{R}/V_1 \\ V_3 &= V_2 \end{aligned} \tag{5.4}$$

(4)
$$\begin{aligned} n &= 4 \\ V_1 &= A_1 \sqrt[8]{R} \\ V_2 &= \sqrt[4]{R} \\ V_3 &= V_2/A_1^2 \\ V_4 &= V_2 \end{aligned} \tag{5.5}$$

where A_1 is the positive root of the following equation [1,5]:

$$1/A_1^2 - A_1^2 - 2(\sqrt[4]{R} - 1)/(\sqrt[4]{R} + 1) = 0$$

For Chebyschev impedance transformers with two, three, and four sections, let us first define the following parameter

$$u_0 = \sin(\pi w/4)$$

where $w = 2\Delta f/f_0$ is the fractional bandwidth [5]. Consequently, for $n = 2$, 3, and 4, we have

(1) $n = 2$

$$V_1^2 = \sqrt{C^2 + R} + C$$
$$V_2 = R/V_1^2$$

where

$$C = (R - 1)u_0^2/[2(2 - u_0^2)],$$

(2) $n = 3$

$$V_1^2 + 2V_1\sqrt{R} - 2\sqrt{R}/V_1 - R/V_1^2 = D$$
$$V_2 = \sqrt{R}/V_1 \tag{5.7}$$
$$V_3 = V_2$$

where

$$D = 3u_0^2(R - 1)/(4 - 3u_0^2)$$

(3) $n = 4$

$$V_1^2 - R(B + \sqrt{B^2 + A^2/R}) = 0$$
$$V_2 = 1/A \tag{5.8}$$
$$V_3 = A^2R/V_1^2$$
$$V_4 = V_2$$

where

$$A^2 = \frac{1 - 1/R}{2t_1t_2} + \sqrt{\left(\frac{1 - 1/R}{2t_1t_2}\right)^2 + \frac{1}{R}}$$
$$B = \frac{1}{2}\left(\frac{A}{A + 1}\right)^2\left[(t_1 + t_2)\left(A^2 - \frac{1}{RA^2}\right) - 2A + \frac{2}{RA}\right]$$
$$t_1 = 2\sqrt{2}/[u_0^2(\sqrt{2} + 1)] - 1$$
$$t_2 = 2\sqrt{2}/[u_0^2(\sqrt{2} - 1)] - 1$$

The quarter-wavelength Chebyschev transformers with five, six, eight, ten, and twelve sections can be designed by using the computer program R-IT given in Section 5.3.

We evaluate the minimum number, n, of sections of the transformer being designed by using one of the following criteria. For maximally flat impedance transformers:

$$n \geqslant \frac{\log|\Gamma_t(R + 1)/(R - 1)|}{\log|\cos[\pi(1 - w/2)/2]|} \tag{5.9}$$

where w is the fractional bandwidth and the permissible reflection coefficient Γ_t is equal to (VSWR − 1)/(VSWR + 1).

Similarly, for the Chebyschev impedance transformers, the number n is evaluated from

$$n \geqslant \frac{\cosh^{-1}|(R - 1)/[\Gamma_t(R + 1)]|}{\cosh^{-1}|1/\cos[\pi(1 - w/2)/2]|} \tag{5.10}$$

The program Q-IT begins by asking for the following input data:

Z_{01}, Z_{02} = characteristic impedances of the lines that are to be matched, in ohms (see Figure 5.1),
f_0 = the center frequency of a required matching band in Hz,
w = the fractional bandwidth ($0 < w < 2$)
VSWR = maximum value of the voltage standing wave ratio in the required matching band,
$B(n)$ = brief name of the maximally flat impedance transformers,
$T(n)$ = brief name of the Chebyschev impedance transformers.

As a result of calculations, we obtain characteristic impedances of the particular sections. After this, we may calculate the geometrical dimensions of these sections when the transformer being designed is constructed of segments of the coaxial line, triplate stripline, or microstrip line. For this purpose, we must run the program Q-IT again by using the command CONTINUE, and introduce the suitable structural data requested by the computer. The flow diagram of the program Q-IT is shown in Figure 5.2.

Check Calculations

Some illustrative results obtained for maximally flat and Chebyschev impedance transformers are given in Table 5.1.

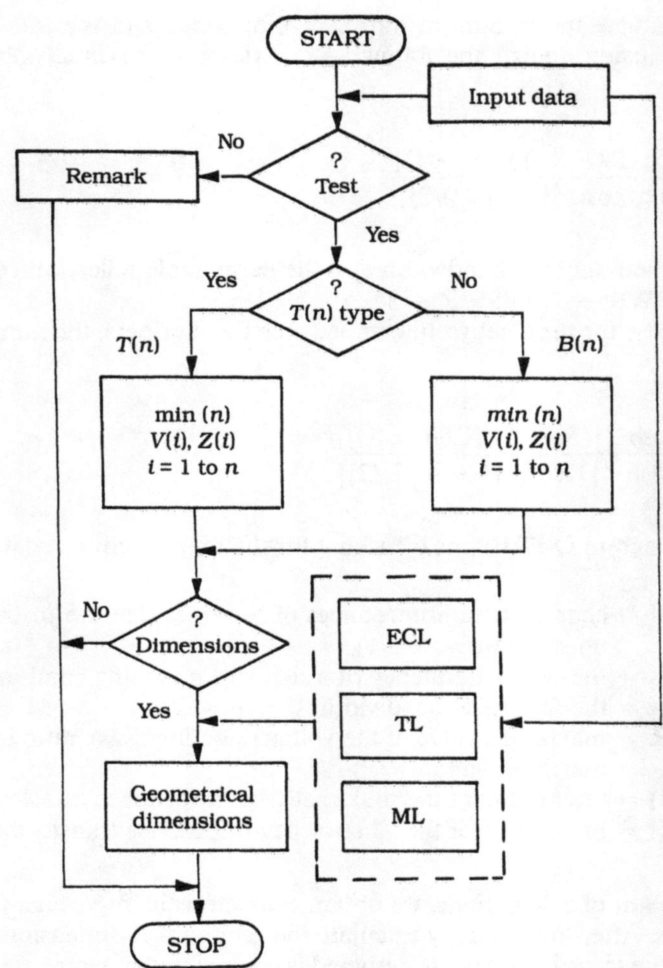

Figure 5.2 The flow diagram of the Q-IT computer program.

Listing of Program Q-IT

```
2500 REM Q-IT
2502 CLS
2504 CHAIN MERGE "TL",2506
2506 CHAIN MERGE "ML",2508
2508 PRINT "QUARTER-WAVELENGTH IMPEDANCE TRANSFORMERS"
```

Table 5.1
Electrical Parameters and Geometrical Dimensions of the Maximally Flat and Chebyschev Response Quarter-Wavelength Impedance Transformers

	(See Figure 5.1)	
Parameters	*B(n)*	*T(n)*
Z_{01}, ohm	25.000	25.000
Z_{02}, ohm	75.000	75.000
f_0, Hz	2 10^9	2 10^9
w	0.500	0.500
VSWR	1.100	1.100
n	3	2
Z_1, ohm	28.698	33.660
Z_2, ohm	43.301	55.703
Z_3, ohm	65.334	—
The transformers are constructed of the air coaxial line segments with the outer conductor diameter $D = 7\ 10^{-3}$ m.		
d_{01}, m	4.615 10^{-3}	4.615 10^{-3}
d_1, m	4.339 10^{-3}	3.994 10^{-3}
d_2, m	3.401 10^{-3}	2.766 10^{-3}
d_3, m	2.356 10^{-3}	—
d_{02}, m	2.005 10^{-3}	2.005 10^{-3}
l_1, m	37.500 10^{-3}	37.500 10^{-3}
l_2, m	37.500 10^{-3}	37.500 10^{-3}
l_3, m	37.500 10^{-3}	—

```
2510 PRINT
2512 PRINT "Data:"
2514 PRINT
2516 INPUT "Zo1[ohm]=",ZO1
2518 INPUT "Zo2[ohm]=",ZO2
2520 INPUT "fo[Hz]=",FO
2522 INPUT "w=",W
2524 INPUT "VSWR=",Q
2526 IF ZO1 <= 0 THEN  PRINT "Error:Zo1 <= 0": GOTO 2514
2528 IF ZO2 <= 0 THEN  PRINT "Error:Zo2 <= 0": GOTO 2514
2530 IF W <= 0 THEN  PRINT "Error:w <= 0": GOTO 2514
2532 IF W >= 2 THEN  PRINT "Error:w >= 2": GOTO 2514
2534 IF Q <= 1 THEN  PRINT "Error:VSWR <= 1": GOTO 2514
2536 DIM Z(4)
2538 LET PI=3.141593
2540 LET G=(Q-1)/(Q+1)
2542 LET U= SIN ( PI *W/4)
2544 IF ZO1 >= ZO2 THEN  LET R=ZO1/ZO2: GOTO 2548
2546 LET R=ZO2/ZO1
2548 PRINT
2550 INPUT "Enter the name ( T(n) or B(n) ) of the insertion lo
ss characteristic:",A$
2552 PRINT
```

```
2554 PRINT "Please wait !"
2556 PRINT
2558 IF A$="T(n)" THEN  GOTO 2566
2560 IF A$="B(n)" THEN  GOTO 2664
2562 PRINT "Wrong command !"
2564 GOTO 2548
2566 REM Procedure T(n)
2568 LET X=(R-1)/(R+1)/G
2570 LET A= LOG(X+ SQR (X*X-1))
2572 LET X=1/( COS ( PI /2*(.99-W/2)))
2574 LET B= LOG(X+ SQR (X*X-1))
2576 LET N= INT (.99+A/B)
2578 IF N=1 THEN  GOTO 2594
2580 IF N=2 THEN  GOTO 2598
2582 IF N=3 THEN  GOTO 2608
2584 IF N=4 THEN  GOTO 2644
2586 PRINT
2588 PRINT "n>4"
2590 PRINT
2592 GOTO 2878
2594 LET Z(1)= SQR (ZO1*ZO2)
2596 GOTO 2708
2598 LET C=(R-1)*U*U/(2*(2-U*U))
2600 LET V= SQR ( SQR (C*C+R)+C)
2602 LET Z(1)=ZO1*V
2604 LET Z(2)=ZO1*R/V
2606 GOTO 2708
2608 LET C=3*U*U*(R-1)/(4-3*U*U)
2610 DEF FN J(V)=V*V+2* SQR (R)*V-2* SQR (R)/V-R/(V*V)-C
2612 LET V1=.01
2614 LET I=1
2616 LET H=.2
2618 IF  FN J(V1)* FN J(V1+I*H) <= 0 THEN  GOTO 2624
2620 LET I=I+1
2622 GOTO 2618
2624 LET V2=V1+I*H
2626 LET V=(V1+V2)/2
2628 IF  ABS ( FN J(V)) <= .000001 THEN  GOTO 2636
2630 IF  FN J(V1)* FN J(V)>0 THEN  LET V1=V: GOTO 2626
2632 LET V2=V
2634 GOTO 2626
2636 LET Z(1)=ZO1*V
2638 LET Z(2)=ZO1* SQR (R)
2640 LET Z(3)=ZO1*R/V
2642 GOTO 2708
2644 LET T1=2* SQR (2)/(( SQR(2)+1)*U*U)-1
2646 LET T2=2* SQR (2)/(( SQR(2)-1)*U*U)-1
2648 LET A=(1-1/R)/(2*T1*T2)+ SQR ((1-1/R)*(1-1/R)/(4*T1*T1*T2*
T2)+1/R)
2650 LET B=.5*A/(( SQR(A)+1)*( SQR(A)+1))*((T1+T2)*(A-1/(A*R))-
2* SQR(A)+2/(R* SQR(A)))
2652 LET V= SQR (R*(B+ SQR (B*B+A/R)))
2654 LET Z(1)=ZO1*V
2656 LET Z(2)=ZO1*V/ SQR(A)
2658 LET Z(3)=ZO1*R/V* SQR(A)
```

```
2660 LET Z(4)=Z01*R/V
2662 GOTO 2708
2664 REM Procedure B(n)
2666 LET A= ABS ((R+1)/(R-1))*G
2668 LET B= ABS ( COS ( PI /2*(1-W/2)))
2670 LET N= INT (.99+ LOG(A)/ LOG(B))
2672 IF N=1 THEN  GOTO 2594
2674 IF N=2 THEN  GOTO 2682
2676 IF N=3 THEN  GOTO 2688
2678 IF N=4 THEN  GOTO 2692
2680 GOTO 2588
2682 LET Z(1)=Z01* SQR ( SQR (R))
2684 LET Z(2)=Z01*R/ SQR ( SQR (R))
2686 GOTO 2708
2688 LET C=0
2690 GOTO 2610
2692 LET S= SQR ( SQR (R))
2694 LET P=2*(S-1)/(S+1)
2696 LET D= SQR (P*P+4)
2698 LET X= SQR ((D-P)/2)
2700 LET Z(1)=Z01*X*R^.125
2702 LET Z(2)=Z01*X*R^.375
2704 LET Z(3)=Z01/X*R^.625
2706 LET Z(4)=Z01/X*R^.875
2708 PRINT "Results, see Fig. 5.1"
2710 PRINT
2712 PRINT "n=";N
2714 PRINT
2716 FOR I=1 TO N
2718 PRINT "Z(";I;")[ohm]=";Z(I)
2720 NEXT I
2722 GOSUB 2888
2724 PRINT
2726 INPUT "Enter the name ( CL ,TL or ML ) of the realization
technique:",B$
2728 PRINT
2730 IF B$="CL" THEN  GOTO 2740
2732 IF B$="TL" THEN  GOTO 2778
2734 IF B$="ML" THEN  GOTO 2828
2736 PRINT "Wrong command !"
2738 GOTO 2724
2740 PRINT "Coaxial line realization"
2742 PRINT
2744 INPUT "D[m]=",D
2746 INPUT "er=",ER
2748 PRINT
2750 PRINT "Geometrical dimensions in meters, see Fig. 5.1(c)"
2752 PRINT
2754 PRINT "do1 =";D/ EXP (Z01* SQR (ER)/60)
2756 PRINT
2758 FOR I=1 TO N
2760 PRINT "d(";I;")=";D/ EXP (Z(I)* SQR (ER)/60)
2762 NEXT I
2764 PRINT
2766 FOR I=1 TO N
```

```
2768 PRINT "l(";I;")=";USING "##.######";7.5E+07/(FO* SQR (ER))
2770 NEXT I
2772 PRINT
2774 PRINT "do2 =";D/ EXP (ZO2* SQR (ER)/60)
2776 GOTO 2878
2778 PRINT "Stripline realization"
2780 PRINT
2782 INPUT "b[m]=",B
2784 INPUT "er=",ER
2786 INPUT "ur=",UR
2788 INPUT "t[m]=",T
2790 PRINT
2792 PRINT "Geometrical dimensions in meters, see Fig. 5.1(b)"
2794 PRINT
2796 LET ZC=ZO1
2798 GOSUB 8252
2800 PRINT
2802 PRINT "Wo1 =";W
2804 FOR G=1 TO N
2806 LET ZC=Z(G)
2808 GOSUB 8252
2810 PRINT
2812 PRINT "W(";G;")=";W
2814 PRINT "l(";G;")=";USING"##.######";7.5E+07/(FO* SQR (ER))
2816 NEXT G
2818 LET ZC=ZO2
2820 GOSUB 8252
2822 PRINT
2824 PRINT "Wo2 =";W
2826 GOTO 2878
2828 PRINT "Microstrip line realization"
2830 PRINT
2832 INPUT "h[m]=",H
2834 INPUT "t[m]=",T
2836 INPUT "er=",ER
2838 PRINT
2840 PRINT "Geometrical dimensions in meters, see Fig. 5.1(b)"
2842 PRINT
2844 LET NN=N
2846 LET F=FO
2848 LET ZC=ZO1
2850 GOSUB 8454
2852 PRINT
2854 PRINT "Wo1 =";(U-DTU)*H
2856 FOR J=1 TO NN
2858 LET ZC=Z(J)
2860 GOSUB 8454
2862 PRINT
2864 PRINT "W(";J;")=";(U-DTU)*H
2866 PRINT "l(";J;")=";USING"##.######";7.5E+07/(FO* SQR (EF))
2868 NEXT J
2870 LET ZC=ZO2
2872 GOSUB 8454
2874 PRINT
2876 PRINT "Wo2 =";(U-DTU)*H
```

```
2878 REM Chain subroutine
2880 PRINT
2882 INPUT"Enter 'CONT':",N$
2884 IF N$="CONT" THEN CLS:CHAIN "MENU",10
2886 GOTO 2880
2888 REM Control subroutine
2890 PRINT
2892 INPUT "Enter 'CONT':",N$
2894 IF N$="CONT" THEN CLS:RETURN
2896 GOTO 2890
```

5.3 THE *R*-TRANSFORMER

A characteristic feature of the Chebyschev impedance transformers described in Section 5.2 is their discontinuity resulting from the properties of the Chebyschev polynomials used in the synthesis. For example, if ratio R of resistances that are to be matched and tolerable reflection coefficient Γ_t are given, then the fractional bandwidth w changes discretely along with the change of the degree n of the polynomial describing the analyzed transformer. This means that for given R and Γ_t we may not always design a quarter-wavelength impedance transformer with a precisely determined bandwidth, because the degree of the Chebyschev polynomials changes by a whole number. From the practical point of view, attempts at designing a stepped impedance transformer fulfilling exactly the predetermined electrical requirements are worth undertaking (see, e.g., [7]).

We [8] can show how this problem can be solved by using the following insertion loss function:

$$L(\theta) = 1 + h^2 T_n^2[(\cos\theta + m)/t] \tag{5.11}$$

where $h = |\Gamma_t|/\sqrt{1 - |\Gamma_t|^2}$ is the attenuation coefficient, m is the synthesis parameter, t is the frequency coefficient, θ is the electrical length and $T_n(x)$ denotes the Chebyschev polynomial of the first kind of degree n. The electrical scheme of the transformer described by (5.11) (termed R-transformer) is shown in Figure 5.3. Obviously, at zero frequency ($\theta = 0$ rad) the attenuation L is

$$L_R = (R + 1)^2/(4R) \tag{5.12}$$

Relations (5.11) and (5.12) therefore imply that:

$$1 + h^2 T_n^2[(1 + m)/t] = (R + 1)^2/(4R)$$

After rearranging the terms, we get

$$T_n[(1 + m)/t] - (R - 1)/(2h\sqrt{R}) = 0 \tag{5.13}$$

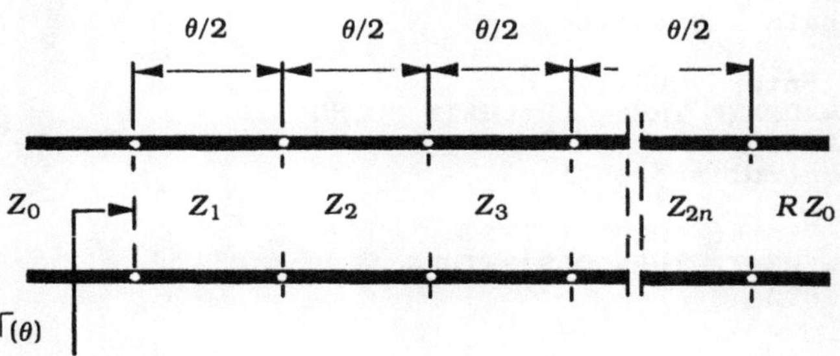

Figure 5.3 The electrical scheme of the $2n$-section R-transformer.

By solving (5.13) for given values of h, n, and R, we obtain the positive root:

$$x_n = (1 + m)/t$$

The first matching band of the transformer under analysis is determined by the angles θ_1 and θ_2 for which:

$$\cos\theta_1 + m = t$$

$$\cos\theta_2 + m = -t$$

Having $x_n = (1 + m)/t$, we can describe this band as follows:

$$-1 \leqslant \frac{\cos\theta + m}{1 + m} x_n \leqslant 1 \tag{5.14}$$

For $0 \leqslant \theta \leqslant 2\pi$, the inequalities in (5.14) are satisfied if

$$\begin{aligned} \theta_1 \leqslant \theta \leqslant 2\pi - \theta_1 \\ 2\pi - \theta_2 \leqslant \theta \leqslant \theta_2 \end{aligned} \tag{5.15}$$

where

$$\theta_1 = \arccos[(1 + m)/x_n - m]$$

$$\theta_2 = \arccos[-(1 + m)/x_n - m]$$

Analysis of (5.14) has shown that the second matching band is constrained by angles $2\pi - \theta_2$ and $2\pi - \theta_1$ (see Figure 5.4). Generally, we may assign such values

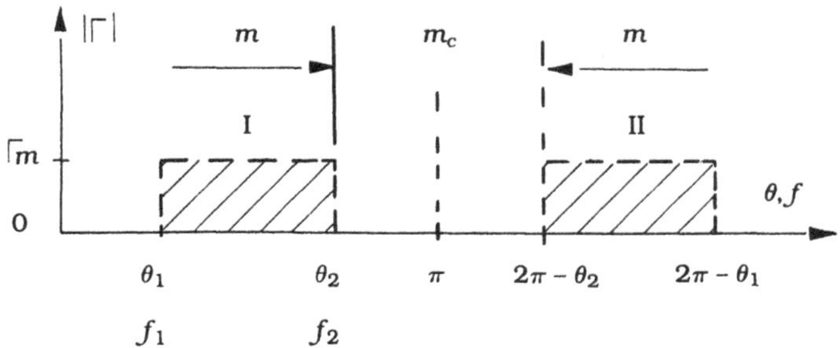

Figure 5.4 Positions of the passbands of the analyzed *R*-transformers.

of $m = m_c$ for which the upper limit of band I overlaps the lower limit of band II. In these extreme cases, $\theta_2 = \pi$, and consequently

$$\arccos[-(1 + m_c)/x_n - m_c] = \pi$$

From the above relations, it follows that

$$m_c = (x_n - 1)/(x_n + 1) \tag{5.16}$$

The complete bandwidth of the adjacent bands 1 and 2 is determined by angles θ_c and $2\pi - \theta_c$, where

$$\theta_c = \arccos[(3 - x_n)/(1 + x_n)] \tag{5.17}$$

Here, we note that the analyzed circuit described by (5.11) is equivalent for $m = m_c$ to the standard Chebyschev transformer of degree $2n$ (i.e., having $2n$ quarter-wavelength sections).

Let us assume now that the required bandwidth w of the transformer being designed is expressed by the coefficient k defined as:

$$k = \theta_2/\theta_1 = (2 + w)/(2 - w) \tag{5.18}$$

Then, from (5.15) and (5.18) we derive the following equation:

$$k \arccos[(1 + m)/x_n - m] - \arccos[-(1 + m)/x_n - m] = 0 \tag{5.19}$$

This equation is solved for $-1 < m \leqslant m_c$. Similarly, the coefficient k must fulfil the condition $1 < k \leqslant \pi/\theta_c$.

Let us suppose now that values of R, Γ_t, and w are given first. For these data, the designing of the R-transformer consists of evaluating such values of n, m and t which satisfy (5.19) and the following constraints:

$$
\begin{aligned}
&n = 1, 2, \ldots, 6 \\
&-1 \leqslant m \leqslant m_c \\
&0 < t < 1
\end{aligned}
\tag{5.20}
$$

This can be done automatically by using the computer program R-IT. As output results, the characteristic impedances Z_i, where $i = 1, 2, \ldots, 2n$, of the particular sections are calculated (see Figure 5.3).

The electrical lengths of these sections at a center frequency $f_0 = (f_1 + f_2)/2$ of the assumed matching band are equal to

$$\theta(f_0)/2 = \arccos(t - m)/(2 - w), \text{ rad}$$

As has been confirmed in the literature [8,9], the R-transformer for $m = 0$ becomes the quarter-wavelength Chebyschev transformer of the same degree n. If $0 < m \leqslant m_c$, the R-transformer has a wider matching band than the quarter-wavelength Chebyschev transformer of the same degree. The widening of the matching band results in an increase of the lengths of the sections. For $m = m_c$, the R-transformer is equivalent to $2n$ degree Chebyschev transformer, that is, having $2n$ quarter-wavelength sections. If $m < 0$, contraction of the matching band takes place as a result of shortening of the electrical length of the sections along with a simultaneous conversion of the R-transformer into a nonmonotonic structure.

Also worth pointing out is that the algorithm presented allows us to design two structurally different R-transformers, monotonic and nonmonotonic, that satisfy the same electrical requirements. This conclusion is illustrated by R-transformers 3 and 4, the main electrical parameters, frequency responses, and structural outlines of which are given in Table 5.2 and in Figures 5.5 and 5.6, respectively.

From a practical standpoint, it is important that the program presented can be used for designing two-, three-, four-, five-, six-, eight-, ten-, and twelve-section quarter-wavelength Chebyschev transformers. For this reason, we must first introduce into the computer memory the value $m = 0$.

The program R-IT begins by introducing into the computer memory the following input data:

Z_{01}, Z_{02} = characteristic impedances of the lines that are to be matched in ohms,

f_0 = the center frequency of a required matching band in Hz,

w = the fractional bandwidth ($0 < w < 2$),
VSWR = maximum value of the voltage standing wave ratio in the required matching band, VSWR = $(1 + |\Gamma_t|)/(1 - |\Gamma_t|)$,
m = 0, for the quarter-wavelength Chebyschev transformers,
m = ?, for the R-transformers.

From calculations, we find the minimal transformer degree n and characteristic impedances Z_i, where $i = 1, 2, \ldots, 2n$, of the individual sections. Moreover, the electrical length $\theta(f_0)/2$ of these sections at center frequency f_0 of the matching band is calculated.

The flow diagram of the program R-IT is presented in Figure 5.7.

Check Calculations

Some typical calculation results obtained by using the computer program R-IT are presented in Table 5.2.

Table 5.2
Main Electrical Parameters of the Four-, Six-, Eight-, Ten- and Twelve-Section R-Transformers

	$Z_{01} = 25\ \Omega$, $Z_{02} = 50\ \Omega$, $VSWR = 1.105263$, $m = ?$				
Parameters	*1*	*2*	*3*	*4*	*5**
n	2	3	4	5	6
w	0.700000	0.950000	1.250000	1.250000	1.511500
Z_1, ohm	29.256	31.373	27.312	35.186	26.719
Z_2, ohm	31.612	25.630	28.510	18.232	27.726
Z_3, ohm	39.542	41.602	31.365	54.665	28.993
Z_4, ohm	42.725	30.046	33.461	16.481	30.522
Z_5, ohm	—	48.769	37.356	68.252	32.299
Z_6, ohm	—	39.843	39.853	18.314	34.293
Z_7, ohm	—	—	43.843	75.843	36.451
Z_8, ohm	—	—	45.767	22.867	38.700
Z_9, ohm	—	—	—	68.560	40.954
Z_{10}, ohm	—	—	—	35.525	43.113
Z_{11}, ohm	—	—	—	—	45.083
Z_{12}, ohm	—	—	—	—	46.782
$\theta(f_0)/2$, rad	0.862377	0.677570	0.849031	0.513546	0.894489

*The R-transformer with adjacent bands. The total fractional bandwidth of this transformer is equal to 1.721825.

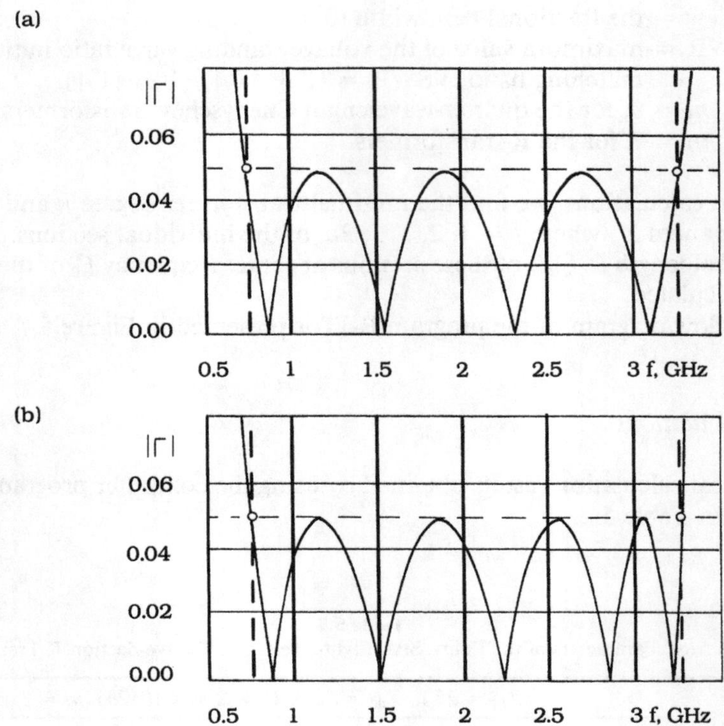

Figure 5.5 Reflection characteristics $|\Gamma(f)|$ of the compared R-transformers: (a) R-transformer 3 and (b) R-transformer 4.

Here, we note that R-transformer 5 is equivalent to the Chebyschev transformer composed of twelve quarter-wavelength sections. In this case, x_n = 1.098588 and $m = m_c = 4.697812 \cdot 10^{-2}$.

Listing of Program R-IT

```
3000 REM R-IT
3002 CLS
3004 PRINT "R-TRANSFORMERS"
3006 CHAIN MERGE "STL",3008
3008 DEFDBL A-Z:DEFINT I,J,N
3010 DEF FNT(V)=4*V*V*V-3*V-TN
3012 DEF FNP(V)=8*V*V*V*V-8*V*V+1-TN
3014 DEF FNQ(V)=16*V*V*V*V*V-20*V*V*V+5*V-TN
3016 DEF FNR(V)=32*V*V*V*V*V*V-48*V*V*V*V+18*V*V-1-TN
```

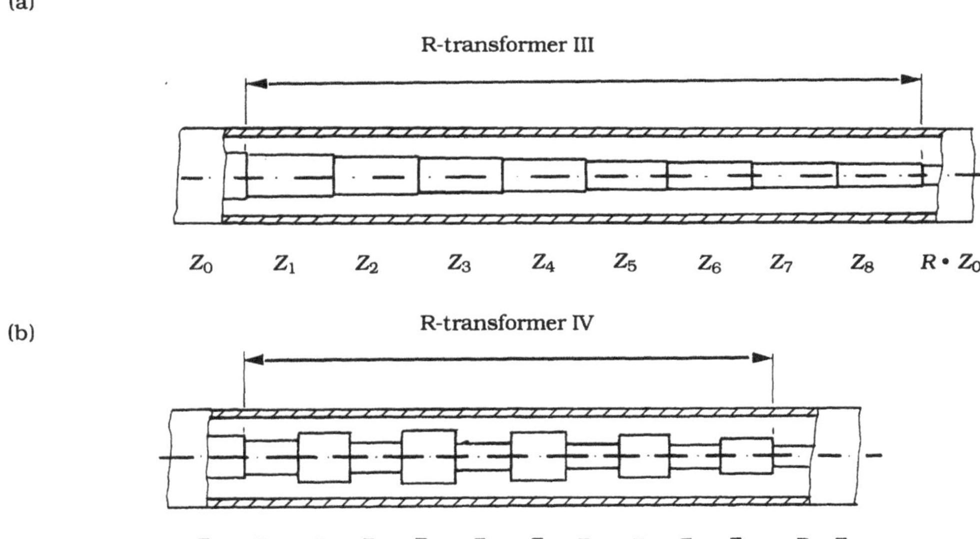

Figure 5.6 Outlines of the compared *R*-transformers: (a) *R*-transformer 3 and (b) *R*-transformer 4.

```
3018 DEF FNW(X)=K*(PI/2-ATN(((1+X)/XN-X)/SQR(1-((1+X)/XN-X)*((1
+X)/XN-X))))-(PI/2-ATN((-(1+X)/XN-X)/SQR(1-(-(1+X)/XN-X)*(-(1+X
)/XN-X))))
3020 PRINT
3022 PRINT "Data:"
3024 PRINT
3026 INPUT "Zo1[ohm]=",ZO1
3028 INPUT "Zo2[ohm]=",ZO2
3030 INPUT "fo[Hz]=",FO
3032 INPUT "w=",WW
3034 INPUT "VSWR=",Q
3036 INPUT "m=",M$
3038 IF ZO1 <= 0 THEN  PRINT "Error:Zo1 <= 0": GOTO 3024
3040 IF ZO2 <= 0 THEN  PRINT "Error:Zo2 <= 0": GOTO 3024
3042 IF ZO2 <= ZO1 THEN  PRINT "Assume  Zo2 >= Zo1": GOTO 3024
3044 IF FO <= 0 THEN  PRINT "Error:fo <= 0": GOTO 3024
3046 IF WW <= 0 THEN  PRINT "Error:w <= 0": GOTO 3024
3048 IF WW >= 2 THEN  PRINT "Error:w >= 2": GOTO 3024
3050 IF Q <= 1 THEN  PRINT "Error: VSWR <= 1": GOTO 3024
3052 LET RI=ZO2/ZO1
3054 LET  PI=3.141592653#
3056 PRINT
3058 PRINT "Please wait !"
3060 PRINT
3062 LET H=(Q-1)/(2* SQR( Q))
```

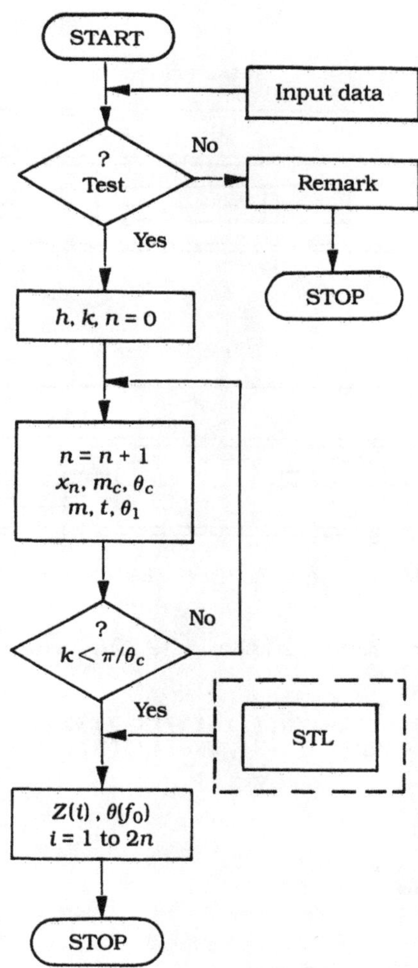

Figure 5.7 The flow diagram of the R-IT computer program.

```
3064 IF (RI-1)/(RI+1) <= (Q-1)/(Q+1) THEN  PRINT "The lines can
 be connected directly !":GOTO 3364
3066 LET QO= ABS ((RI-1)/(RI+1))
3068 LET K=(2+WW)/(2-WW)
3070 LET TN=QO/(H* SQR (1-QO*QO))
3072 IF M$="0" THEN  GOTO 3084
3074 IF M$="?" THEN  GOTO 3116
3076 PRINT
3078 PRINT "Wrong command !"
```

```
3080 PRINT
3082 GOTO 3036
3084 LET M=0
3086 LET X=(RI-1)*(Q+1)/((RI+1)*(Q-1))
3088 LET A= LOG(X+ SQR (X*X-1))
3090 LET X=1/ COS ( PI /2*(.99-WW/2))
3092 LET B= LOG(X+ SQR (X*X-1))
3094 LET N= INT (.99+A/B)
3096 IF N=1 THEN  LET FL=1:LET PN=1: GOTO 3124
3098 IF N=2 THEN  LET FL=1:LET PN=1: GOTO 3132
3100 IF N=3 THEN  LET FL=1:LET PN=1: GOTO 3140
3102 IF N=4 THEN  LET FL=1:LET PN=1: GOTO 3172
3104 IF N=5 THEN  LET FL=1:LET PN=1: GOTO 3206
3106 IF N=6 THEN  LET FL=1:LET PN=1: GOTO 3240
3108 IF N>12 THEN  PRINT "n>12": GOTO 3364
3110 IF N>10 THEN  LET FL=2:LET PN=1: GOTO 3238
3112 IF N>8 THEN  LET FL=2:LET PN=1: GOTO 3204
3114 IF N>6 THEN  LET FL=2:LET PN=1: GOTO 3170
3116 REM Procedure R-IT
3118 LET FL=0
3120 LET N=1
3122 LET PN=1
3124 LET XN=TN
3126 LET MC=(3-XN)/(1+XN)
3128 IF K <=  PI /(PI/2-ATN(MC/SQR(1-MC*MC))) THEN  GOTO 3276
3130 LET N=2
3132 LET XN= SQR ((TN+1)/2)
3134 LET MC=(3-XN)/(1+XN)
3136 IF K <=  PI /(PI/2-ATN(MC/SQR(1-MC*MC))) THEN  GOTO 3276
3138 LET N=3
3140 LET V1=.01
3142 LET J=1
3144 LET DV=.2
3146 IF  FNT(V1)* FNT(V1+J*DV) <= 0 THEN  GOTO 3152
3148 LET J=J+1
3150 GOTO 3146
3152 LET V2=V1+J*DV
3154 LET V=(V1+V2)/2
3156 IF  ABS (FNT(V))<= .000001 THEN  GOTO 3164
3158 IF  FNT(V1)* FNT(V)>0 THEN  LET V1=V: GOTO 3154
3160 LET V2=V
3162 GOTO 3154
3164 LET XN=V
3166 LET MC=(3-XN)/(1+XN)
3168 IF K <=  PI /(PI/2-ATN(MC/SQR(1-MC*MC))) THEN  GOTO 3276
3170 LET N=4
3172 LET V1=.01
3174 LET J=1
3176 LET DV=.2
3178 IF  FNP(V1)* FNP(V1+J*DV) <= 0 THEN  GOTO 3184
3180 LET J=J+1
3182 GOTO 3178
3184 LET V2=V1+J*DV
3186 LET V=(V1+V2)/2
3188 IF  ABS(FNP(V))<= .000001 THEN  GOTO 3196
```

```
3190 IF  FNP(V1)* FNP(V)>0 THEN  LET V1=V: GOTO 3186
3192 LET V2=V
3194 GOTO 3186
3196 LET XN=V
3198 IF FL=2 THEN  GOTO 3276
3200 LET MC=(3-XN)/(1+XN)
3202 IF K <=  PI /(PI/2-ATN(MC/SQR(1-MC*MC))) THEN  GOTO 3276
3204 LET N=5
3206 LET V1=.01
3208 LET J=1
3210 LET DV=.2
3212 IF  FNQ(V1)* FNQ(V1+J*DV) <= 0 THEN  GOTO 3218
3214 LET J=J+1
3216 GOTO 3212
3218 LET V2=V1+J*DV
3220 LET V=(V1+V2)/2
3222 IF  ABS( FNQ(V))<= .000001 THEN  GOTO 3230
3224 IF  FNQ(V1)* FNQ(V)>0 THEN  LET V1=V: GOTO 3220
3226 LET V2=V
3228 GOTO 3220
3230 LET XN=V
3232 IF FL=2 THEN  GOTO 3276
3234 LET MC=(3-XN)/(1+XN)
3236 IF K <=  PI /(PI/2-ATN(MC/SQR(1-MC*MC))) THEN  GOTO 3276
3238 LET N=6
3240 LET V1=.01
3242 LET J=1
3244 LET DV=.2
3246 IF  FNR(V1)* FNR(V1+J*DV) <= 0 THEN  GOTO 3252
3248 LET J=J+1
3250 GOTO 3246
3252 LET V2=V1+J*DV
3254 LET V=(V1+V2)/2
3256 IF  ABS( FNR(V))<= .000001 THEN  GOTO 3264
3258 IF  FNR(V1)* FNR(V)>0 THEN  LET V1=V: GOTO 3254
3260 LET V2=V
3262 GOTO 3254
3264 LET XN=V
3266 IF FL=2 THEN  GOTO 3276
3268 LET MC=(3-XN)/(1+XN)
3270 IF K <=  PI /(PI/2-ATN(MC/SQR(1-MC*MC))) THEN  GOTO 3276
3272 PRINT "n>6"
3274 GOTO 3364
3276 IF FL=0 THEN  GOTO 3282
3278 IF FL=1 THEN  GOTO 3312
3280 IF FL=2 THEN  LET M=(XN-1)/(XN+1): GOTO 3312
3282 LET X1=.99*(XN-1)/(XN+1)
3284 LET J=1
3286 LET DX=.01
3288 LET X2=X1-J*DX
3290 IF X2 <= -.99 THEN PRINT"Unfortunately  m<=.990": GOTO 336
4
3292 IF (FNW(X1)*FNW(X2)) <0 THEN  GOTO 3298
3294 LET J=J+1
3296 GOTO 3288
```

```
3298 LET X2=X1-J*DX
3300 LET X=(X1+X2)/2
3302 IF  ABS( FNW(X))<= .000001 THEN  GOTO 3310
3304 IF  FNW(X1)*FNW(X)>0 THEN  LET X1=X: GOTO 3300
3306 LET X2=X
3308 GOTO 3300
3310 LET M=X
3312 LET T=(1+M)/XN
3314 IF PN=1 THEN GOSUB 1688
3316 IF PN=2 THEN LET D=M-1:LET C=1/SQR(RI):GOSUB 1710
3318 PRINT
3320 PRINT "Results, see Fig. 5.3"
3322 PRINT
3324 IF FL=0 THEN PRINT "n=";N
3326 PRINT
3328 FOR I=1 TO 2*N
3330 PRINT "Z(";I;")[ohm]=";USING "###.#######";Z01*Z(2*N+1-I)
3332 NEXT I
3334 PRINT
3336 IF FL>.1 THEN GOTO 3342
3338 PRINT "O(fo)/2[rad]=";USING "##.#######";1/(2-WW)*(PI/2-ATN
((T-M)/SQR(1-(T-M)*(T-M))))
3340 GOSUB 3374
3342 IF FL=1 THEN  PRINT "O(fo)/2[rad]=PI/4": GOTO 3364
3344 IF FL=2 THEN  PRINT "O(fo)/2[rad]=PI/2": GOTO 3364
3346 IF N >6 THEN GOTO 3364
3348 IF PN>2 THEN GOTO 3364
3350 LET N=N+1
3352 LET PN=PN+1
3354 IF N=2 THEN GOTO 3130
3356 IF N=3 THEN GOTO 3138
3358 IF N=4 THEN GOTO 3170
3360 IF N=5 THEN GOTO 3204
3362 IF N=6 THEN GOTO 3238
3364 REM Chain subroutine
3366 PRINT
3368 INPUT"Enter 'CONT':",N$
3370 IF N$="CONT" THEN CLS:CHAIN "MENU",10
3372 GOTO 3366
3374 REM Control subroutine
3376 PRINT
3378 INPUT "Enter 'CONT':",N$
3380 IF N$="CONT" THEN CLS:RETURN
3382 GOTO 3376
```

5.4 MICROWAVE DIRECTIONAL COUPLERS

This section presents four algorithms for designing common planar TEM-mode directional couplers, namely:

(1) 3/2 wavelength hybrid rings denoted as DC/I,
(2) two-branch line couplers denoted as DC/II,

(3) one-section coupled transmission line couplers denoted as DC/III and
(4) asymmetric Chebyschev high-pass couplers denoted as DC/IV if their mean couplings are equal to 8.34, 10 or 20 dB.

The electrical diagrams and structural outlines of these circuits are shown in Figures 5.8, 5.9, 5.10 and 5.11, respectively [5,6,10,11,13].

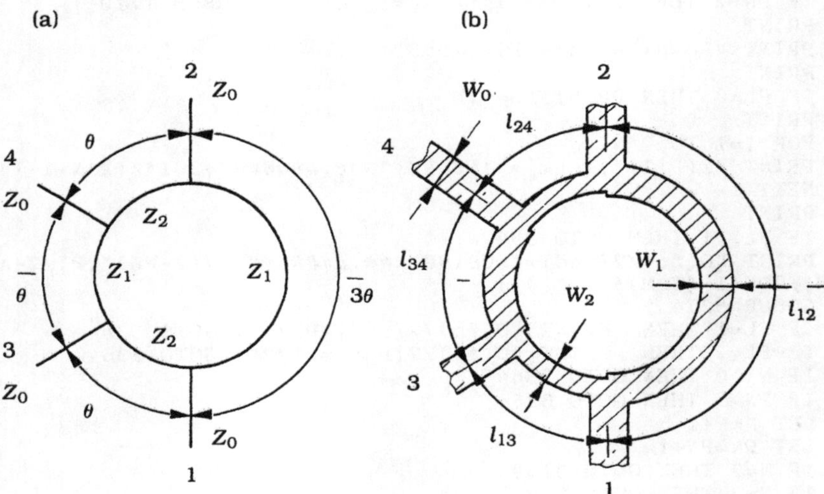

Figure 5.8 The hybrid ring directional coupler: (a) electrical diagram and (b) microstrip line configuration.

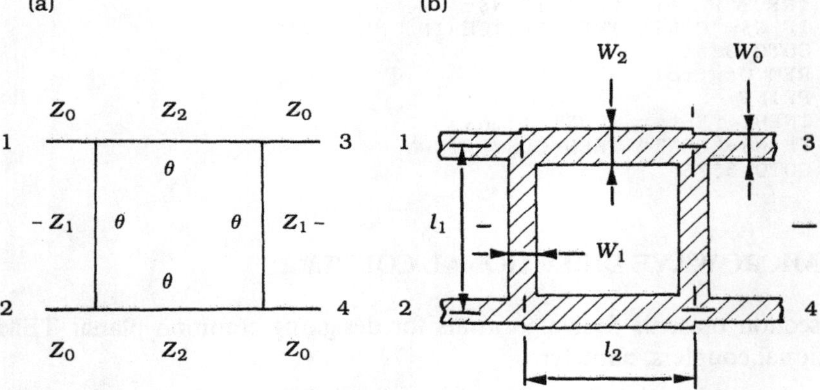

Figure 5.9 The two-branch line directional coupler: (a) electrical diagram and (b) microstrip line configuration.

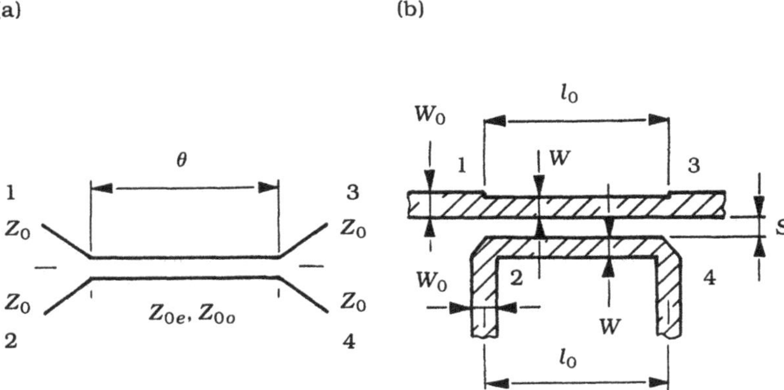

Figure 5.10 The one-section coupled transmission line directional coupler: (a) electrical diagram and (b) the coupled microstrip lines configuration.

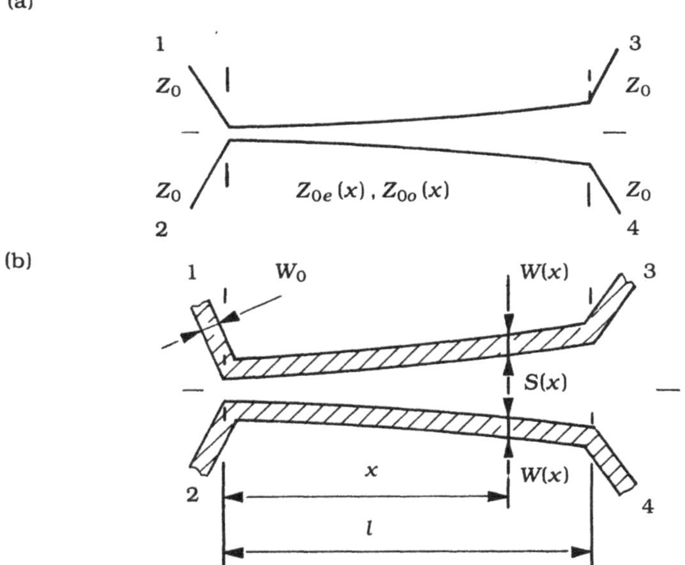

Figure 5.11 The high-pass transmission line directional coupler: (a) electrical scheme and (b) the coupled microstrip lines configuration.

Let us assume that all four ports of the hybrid ring shown in Figure 5.8 are terminated in real impedances Z_0. Under this assumption, the scattering parameters of this coupler, obtained by means of even and odd mode excitation method, are

$$\begin{aligned}
S_{11} &= S_{22} = (S_{11}^{+} + S_{11}^{-})/2 \\
S_{33} &= S_{44} = (S_{22}^{+} + S_{22}^{-})/2 \\
S_{12} &= S_{21} = (S_{11}^{+} - S_{11}^{-})/2 \\
S_{34} &= S_{43} = (S_{22}^{+} - S_{22}^{-})/2 \\
S_{13} &= S_{31} = S_{24} = S_{42} = (S_{12}^{+} + S_{12}^{-})/2 \\
S_{14} &= S_{41} = S_{23} = S_{32} = (S_{12}^{+} - S_{12}^{-})/2
\end{aligned} \tag{5.21}$$

where

$$\begin{aligned}
S_{11}^{+} &= (1 - A - B - jD)/[1 + A + B + j(C + E)] \\
S_{11}^{-} &= (1 - A' + B' + jD')/[1 + A' - B' - j(C' - E)] \\
S_{22}^{+} &= (1 - A - B + jD)/[1 + A + B + j(C + E)] \\
S_{22}^{-} &= (1 - A' + B' - jD')/[1 + A' - B' - j(C' - E)] \\
S_{12}^{+} &= -j2y_2/\{\sin(\theta)[1 + A + B + j(C + E)]\} \\
S_{12}^{-} &= -j2y_2/\{\sin(\theta)[1 + A' - B' - j(C' - E)]\} \\
A &= y_2^2 - y_1^2 \tan(k\theta/2)\tan(\theta/2) \\
A' &= y_2^2 - y_1^2 \cot(k\theta/2)\cot(\theta/2) \\
B &= y_1 y_2 \cot(\theta)[\tan(k\theta/2) + \tan(\theta/2)] \\
B' &= y_1 y_2 \cot(\theta)[\cot(k\theta/2) + \cot(\theta/2)] \\
C &= y_1[\tan(k\theta/2) + \tan(\theta/2)] \\
C' &= y_1[\cot(k\theta/2) + \cot(\theta/2)] \\
D &= y_1[\tan(k\theta/2) - \tan(\theta/2)] \\
D' &= y_1[\cot(k\theta/2) - \cot(\theta/2)] \\
E &= -2y_2 \cot(\theta) \\
j &= \sqrt{-1} \\
k &= 3 \\
y_1 &= Z_0/Z_1 \quad \text{and} \quad y_2 = Z_0/Z_2.
\end{aligned}$$

At center band frequency f_0, for which $\theta = \pi/2$ radians, these scattering parameters become:

$$\begin{aligned} S_{11} &= S_{22} = S_{33} = S_{44} = (1 - y_1^2 - y_2^2)/V \\ S_{12} &= S_{21} = -S_{34} = -S_{43} = j2y_1/V \\ S_{13} &= S_{31} = S_{24} = S_{42} = -j2y_2/V \\ S_{14} &= S_{41} = S_{23} = S_{32} = 0 \end{aligned} \tag{5.22}$$

where $V = 1 + y_1^2 + y_2^2$. From (5.22), it follows that all four ports are perfectly matched ($S_{11} = S_{22} = S_{33} = S_{44} = 0$) at the center frequency f_0 if

$$y_1^2 + y_2^2 = 1 \tag{5.23}$$

For simplicity, let us assume that the coupler described in (5.22) and (5.23) is fed at the arm 4 (see Figure 5.8). In this case, the input power is fully divided between ports 2 and 3, and port 1 is separated. Because the physical distances to the adjacent ports 2 and 3 are equal, the phase shift of signals at these ports is equal to 0. The characteristic admittances of the ring sections are calculated from condition (5.23), and the following relation:

$$C_{34} = 20 \log(1/|S_{34}|) = 10 \log(1/y_1^2), \text{ dB} \tag{5.24}$$

where C_{34} denotes the coupling between ports 3 and 4 at $\theta = \pi/2$ radians. The response of input VSWR and isolation, I, are calculated from (5.21) as follows:

$$\begin{aligned} \text{VSWR}(\theta) &= [1 + |S_{44}(\theta)|]/[1 - |S_{44}(\theta)|] \\ I(\theta) &= 20 \log[1/|S_{14}(\theta)|], \text{ dB} \end{aligned} \tag{5.25}$$

When port 1 is fed, the input power is divided between ports 2 and 3, and port 4 is separated. The phase shift between output signals at center frequency f_0 is equal to π radians. The coupling response between ports 1 and 3 is given by:

$$C_{31}(\theta) = 20 \log[1/|S_{31}(\theta)|], \text{ dB} \tag{5.26}$$

From (5.22) and (5.23), it follows that for $\theta = \pi/2$ radians, the coupling $C_{31}(\theta)$ becomes:

$$C_{31} = 10 \log(1/y_2^2), \text{ dB} \tag{5.27}$$

In this case, the input VSWR and isolation I are:

$$\begin{aligned} \mathrm{VSWR}(\theta) &= [1 + |S_{11}(\theta)|]/[1 - |S_{11}(\theta)|] \\ I(\theta) &= 20\log[1/|S_{41}(\theta)|], \text{ dB} \end{aligned} \tag{5.28}$$

It is relatively simple to prove that for $k = 1$ (see (5.21)), the coupler under analysis transforms into the simpler structure shown in Figure 5.9. In the literature the directional coupler thus obtained is called a two-branch hybrid [10]. Its scattering parameters, derived from (5.21) at $\theta = \pi/2$ radians, are:

$$\begin{aligned} S_{11} &= S_{22} = S_{33} = S_{44} = (1 - y_1^2 + y_2^2)(1 + y_1^2 - y_2^2)/W \\ S_{12} &= S_{21} = S_{34} = S_{43} = -j2y_1(1 + y_1^2 - y_2^2)/W \\ S_{13} &= S_{31} = S_{24} = S_{42} = -j2y_2(1 + y_2^2 - y_1^2)/W \\ S_{14} &= S_{41} = S_{23} = S_{32} = -4y_1y_2/W \end{aligned} \tag{5.29}$$

where $W = (1 + y_2^2 - y_1^2)^2 + 4y_1^2$. From (5.29), two matching conditions follow, for which $S_{11} = S_{22} = S_{33} = S_{44} = 0$. Let us assume that the second, namely

$$y_2^2 - y_1^2 = 1 \tag{5.30}$$

is satisfied. Under this assumption the parameters of (5.29) reduce to the following form:

$$\begin{aligned} S_{11} &= S_{22} = S_{33} = S_{44} = 0 \\ S_{12} &= S_{21} = S_{34} = S_{43} = 0 \\ S_{13} &= S_{31} = S_{24} = S_{42} = -j/y_2 \\ S_{14} &= S_{41} = S_{23} = S_{32} = -y_1/y_2 \end{aligned} \tag{5.31}$$

The above parameters illustrate how an input power delivered to port 1 (see Figure 5.9), is divided between output ports 3 and 4. We note that the phases of the output signals are shifted by $\pi/2$ radians. The frequency responses of input VSWR, coupling C_{41} and isolation I are calculated from the following formulas:

$$\begin{aligned} \mathrm{VSWR}(\theta) &= [1 + |S_{11}(\theta)|]/[1 - |S_{11}(\theta)|] \\ C_{41}(\theta) &= 20\log[1/|S_{41}(\theta)|], \text{ dB} \\ I(\theta) &= 20\log[1/|S_{21}(\theta)|], \text{ dB} \end{aligned} \tag{5.32}$$

where $S_{11}(\theta)$, $S_{21}(\theta)$ and $S_{41}(\theta)$ are calculated by using (5.21) for $k = 1$ and $y_2^2 - y_1^2 = 1$.

For $\theta = \pi/2$ radians, the coupling C_{41} takes its nominal value:

$$C_{41} = 10 \log[(1 + y_1^2)/y_1^2], \text{ dB} \tag{5.33}$$

In practice, (5.30) and (5.33) are most frequently utilized for design. Here, we point out that both the directional couplers presented above are narrow-band ones. Nevertheless, their performances are quite adequate for many applications in which bandwidths of up to 10 percent are required [10].

Now, let us consider the wider band directional coupler whose electrical scheme and constructional layout are shown in Figure 5.10. The frequency performance of this type of coupler results directly from the following scattering matrix (see e.g. [5,10,11]):

$$[S] = \begin{bmatrix} 0 & S_{12} & S_{13} & 0 \\ S_{12} & 0 & 0 & S_{13} \\ S_{13} & 0 & 0 & S_{12} \\ 0 & S_{13} & S_{12} & 0 \end{bmatrix} \tag{5.34}$$

where

$$S_{12} = \frac{jk \sin(\theta)}{\sqrt{1 - k^2} \cos(\theta) + j \sin(\theta)}$$

$$S_{13} = \frac{\sqrt{1 - k^2}}{\sqrt{1 - k^2} \cos(\theta) + j \sin(\theta)}$$

$$k = (Z_{0e} - Z_{0o})/(Z_{0e} + Z_{0o})$$

Z_{0e} = the even-mode characteristic impedance, and Z_{0o} = the odd-mode characteristic impedance.

According to (5.34), the coupling response $C_{21}(\theta)$ may be written as

$$C_{21}(\theta) = 10 \log \left[\frac{1 + (1 - k^2) \cot^2(\theta)}{k^2} \right], \text{ dB} \tag{5.35}$$

At the center frequency f_0, for which $\theta = \pi/2$ radians, the coupling $C_{21}(\theta)$ achieves its minimal value:

$$C_{21} = 10 \log(1/k^2), \text{ dB} \tag{5.36}$$

The design of this coupler type is very simple and consists of evaluating the characteristic impedances Z_{0e} and Z_{0o} by using the formulas

$$Z_{0e} = Z_0\sqrt{(1 + k)/(1 - k)}$$
$$Z_{0o} = Z_0\sqrt{(1 - k)/(1 + k)} \quad (5.37)$$

where Z_0 denotes the terminating real impedances, and k is the coupling coefficient obtained from (5.36). As we have mentioned earlier, the electrical length θ of this directional coupler is equal to $\pi/2$ radians at the center band frequency. The usuable bandwidth of this directional coupler is approximately one octave [12].

Many applications, however, require the use of directional couplers with very broad bands. In these cases, the asymmetric high-pass Chebyschev directional coupler, presented in Figure 5.11, can be used. It consists of two transmission lines which are nonuniformly coupled. The coupling factor $k(x)$ changes continuously along the section l. Usually, it is defined in terms of even- and odd-mode characteristic impedances Z_{0e} and Z_{0o} as

$$k(x) = [Z_{0e}(x) - Z_{0o}(x)]/[Z_{0e}(x) + Z_{0o}(x)] \quad (5.38)$$

We assume that ports 1 to 4 of the coupler terminate in two real impedances Z_0 (e.g., 50 Ω) and that the matching condition

$$Z_{0e}(x)Z_{0o}(x) = Z_0^2 \quad (5.39)$$

holds along the whole coupling section l (i.e., for $0 \leqslant x \leqslant l$). This transmission line directional coupler has, under condition (5.39), theoretically perfect input VSWR and isolation I (between ports 1 and 4, and 2 and 3) at all frequencies. In the literature [13], Arndt has shown that the coupling response $C_{21}(f)$, similar to that shown in Figure 5.12, can be achieved by choice of the suitable function $k(x)$. For this process, the approach and design data given in [13] can be used. Generally,

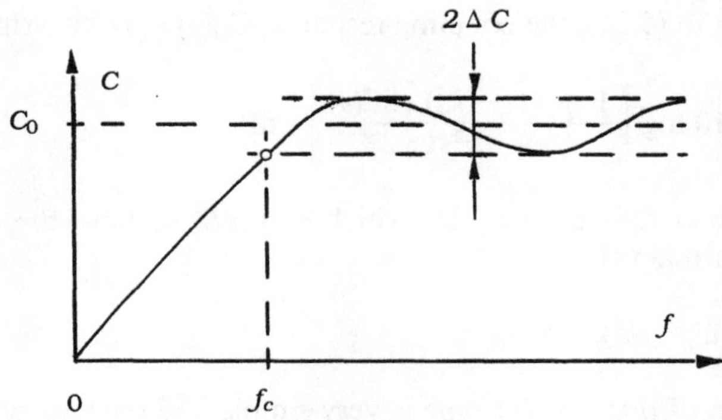

Figure 5.12 Magnitude of the typical coupling response.

the design of this coupler type starts by evaluating the function $k(x/l)$ corresponding to the assumed technical requirements,

$$k(x/l) = \sum_{m=0}^{6} k_m(x/l)^m \tag{5.40}$$

where $0 \leqslant x/l \leqslant 1$, l is the geometrical length of the coupling section and k_0, k_1, k_2, . . . , k_6 are the design coefficients derived by Arndt [13]. These coefficients are related, of course, to a mean coupling C and tolerable coupling deviation ΔC. The values of these coefficients for couplings 8.34, 10 and 20 dB and 1% voltage ripple are given in Table 5.3. The geometrical length l of the coupling section is determined by the low-frequency cut-off coefficient $(l/\lambda)_c$, the values of which are also given in Table 5.3, and by the desired actual maximum wavelength λ_{max}. Thus,

$$l = \lambda_{max}(l/\lambda)_c \tag{5.41}$$

Table 5.3
Coefficients of the Coupling Factor $k(x/l)$ Given by [13]

C,	*Voltage Ripple — 1%*							
dB	k_0	k_1	k_2	k_3	k_4	k_5	k_6	$(l/\lambda)_c$
8.34	0.6688	−0.9751	−0.6127	1.6194	−0.4024	−0.5828	0.2938	0.354
10.00	0.5749	−0.9500	−0.2495	1.2963	−0.6825	−0.0996	0.1174	0.358
20.00	0.1980	−0.4121	0.1856	0.1304	−0.1095	−0.0044	0.0141	0.365

Finally, the characteristic impedances $Z_{0e}(x_i/l)$ and $Z_{0o}(x_i/l)$ for discrete values of x_i/l are calculated from the following standard formulas:

$$\begin{aligned} Z_{0e}(x_i/l) &= Z_0\sqrt{[1 + k(x_i/l)]/[1 - k(x_i/l)]} \\ Z_{0o}(x_i/l) &= Z_0^2/Z_{0e}(x_i/l) \end{aligned} \tag{5.42}$$

All mathematical relationships presented in this section constitute the theoretical basis for the computer program DC. The input data for this program are:

DC/x = the suitable name of the directional coupler that is to be designed (see points (1), (2), (3), and (4) given at the beginning of this section),
Z_0 = the terminating impedances in ohms,
C = the mean value of the coupling in dB,
f = the operating frequency in Hz,
f_0 = the center band frequency in Hz,
υ = the phase shift between output signals in rad,

f_c = the cut-off frequency in Hz, when the asymmetric Chebyschev high-pass coupler is designed.

Here, we note that the phase shift $v(f_0)$ takes only three discrete values (i.e., 0, $\pi/2$ and π radians). Moreover, the DC program makes it possible to design the asymmetric Chebyschev high-pass directional couplers only for three discrete values of mean coupling *C*, 8.34, 10 and 20 dB.

After introducing the above input data into the computer memory, the suitable conditions of physical realization are checked. If these conditions are not satisfied, then an appropriate report is printed, and, of course, the design process is discontinued. Let us suppose that all input data are acceptable for design. The calculations are then made in accordance with the flow diagram shown in Figure 5.13.

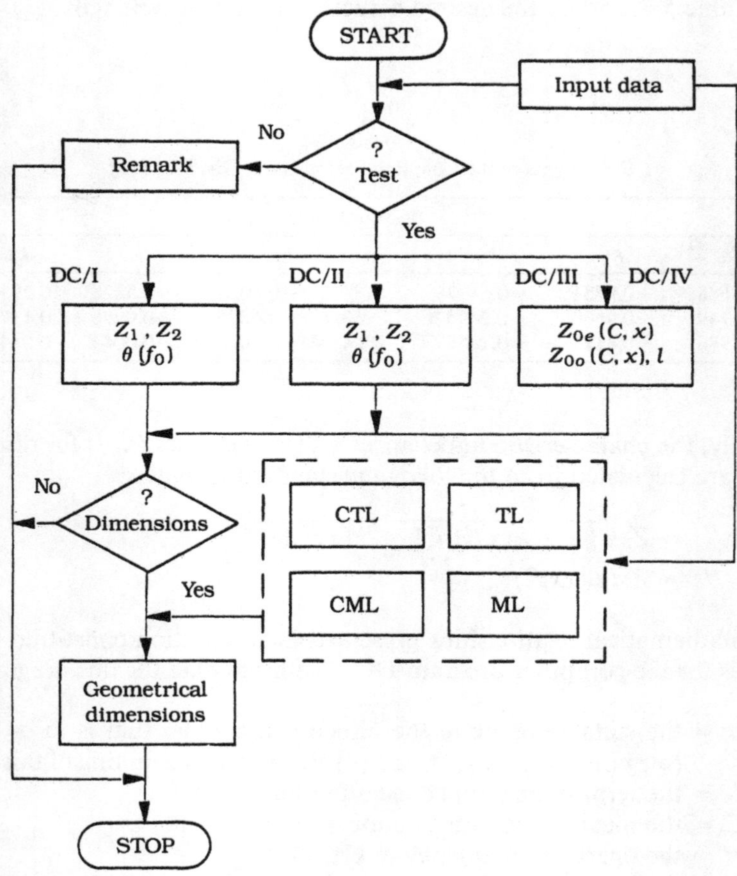

Figure 5.13 The flow diagram of the DC computer program.

At this stage of design, we obtain as the output results the main electrical parameters of the coupler being designed, as well as the input VSWR and coupling, C, at the given operating frequency, f. Next, the geometrical dimensions of the designed coupler can be evaluated. For this purpose, we have to run the DC program again, by means of command CONTINUE, and introduce the suitable values of structural data requested by the computer.

According to Figure 5.13, at this stage of design, some synthesis programs presented in Chapter 6 are called out as subroutines.

Check Calculations

The manner of using the DC computer program for designing the directional couplers discussed above is illustrated by the following computation example.

Input Data

DC/III

$Z_0 = 50,$ Ω

$C = 13,$ dB

$f = 1.3\ 10^9,$ Hz

$f_0 = 1\ 10^9,$ Hz

$v = \pi/2,$ rad

Results

$Z_{0e} = 62.787,$ Ω

$Z_{0o} = 39.817,$ Ω

$\theta(f_0) = \pi/2,$ rad

$\mathrm{VSWR}(f) = 1,$

$C(f) = 13.96,$ dB

Let us assume that the coupler is constructed of segments of the single and coupled triplate striplines with $\epsilon_r = 2.55$, $\mu_r = 1$, $b = 3.52\ 10^{-3}$ m and $t \ll b$. For these data, the geometrical dimensions of the coupler being designed are: $W_0 = 2.601\ 10^{-3}$ m, $W = 2.344\ 10^{-3}$ m, $S = 0.408\ 10^{-3}$ m and $l = 46.96\ 10^{-3}$ m (see Figure 5.10(b)).

Listing of Program DC

```
3500 REM DC
3502 CLS
3504 CHAIN MERGE "ML",3506
3506 CHAIN MERGE "TL",3508
3508 CHAIN MERGE "CTL",3510
3510 PRINT "DIRECTIONAL COUPLERS"
3512 PRINT
3514 PRINT "Data:"
3516 PRINT
3518 INPUT "Enter the name ( DC/I, DC/II, DC/III or DC/IV ) of
```

```
the coupler to be designed:",A$
3520 PRINT
3522 LET PI=3.141593
3524 IF A$="DC/I" THEN  LET K=3: GOTO 3536
3526 IF A$="DC/II" THEN  LET K=1: GOTO 3536
3528 IF A$="DC/III" THEN  GOTO 3536
3530 IF A$="DC/IV" THEN  GOTO 3710
3532 PRINT "Wrong command !"
3534 GOTO 3516
3536 INPUT "Zo[ohm]=",ZO
3538 INPUT "C[dB]=",CO
3540 INPUT "f[Hz]=",F
3542 INPUT "fo[Hz]=",FO
3544 IF ZO <= 0 THEN  PRINT "Error:Zo <= 0": GOTO 3520
3546 IF CO <= 0 THEN  PRINT "Error:C <= 0": GOTO 3520
3548 IF F <= 0 THEN  PRINT "Error:f <= 0": GOTO 3520
3550 IF FO <= 0 THEN  PRINT "Error:fo <= 0": GOTO 3520
3552 IF A$="DC/II" THEN  GOTO 3632
3554 IF A$="DC/III" THEN  GOTO 3670
3556 INPUT "v[rad]=",V$
3558 IF V$="0" THEN  GOTO 3564
3560 IF V$="PI" THEN  GOTO 3598
3562 PRINT "The phase shift v takes only values equal to 0 and
PI radians": GOTO 3556
3564 PRINT
3566 PRINT "Results, see Fig. 5.8"
3568 PRINT
3570 PRINT "4-input port"
3572 PRINT "3-coupled port"
3574 PRINT "1-isolated port"
3576 PRINT
3578 LET Y1= SQR (1/(10^(CO/10)))
3580 LET Y2= SQR (1-Y1*Y1)
3582 PRINT "Z1[ohm]=";ZO/Y1
3584 PRINT "Z2[ohm]=";ZO/Y2
3586 PRINT "O(fo)[rad]= PI/2"
3588 GOSUB 4060
3590 PRINT "VSWR(f)=";(1+M33)/(1-M33)
3592 PRINT "C(f)[dB]=";20/ LOG(10)*LOG(1/M34)
3594 GOSUB 4232
3596 GOTO 3802
3598 PRINT
3600 PRINT "Results, see Fig. 5.8"
3602 PRINT
3604 PRINT "1-input port"
3606 PRINT "3-coupled port"
3608 PRINT "4-isolated port"
3610 PRINT
3612 LET Y2= SQR (1/(10^(CO/10)))
3614 LET Y1= SQR (1-Y2*Y2)
3616 PRINT "Z1[ohm]=";ZO/Y1
3618 PRINT "Z2[ohm]=";ZO/Y2
3620 PRINT "O(fo)[rad]= PI/2"
3622 GOSUB 4060
3624 PRINT "VSWR(f)=";(1+M11)/(1-M11)
```

```
3626 PRINT "C(f)[dB]=";20/ LOG(10)*LOG(1/M13)
3628 GOSUB 4232
3630 GOTO 3802
3632 PRINT
3634 PRINT "Results, see Fig. 5.9"
3636 PRINT
3638 PRINT "1-input port"
3640 PRINT "4-coupled port"
3642 PRINT "2-isolated port"
3644 PRINT
3646 LET AT=10^(CO/10)
3648 LET Y1= SQR (1/(AT-1))
3650 LET Y2= SQR (1+Y1*Y1)
3652 PRINT "Z1[ohm]=";ZO/Y1
3654 PRINT "Z2[ohm]=";ZO/Y2
3656 PRINT "O(fo)[rad]= PI/2"
3658 PRINT "v(fo)[rad]= PI/2"
3660 GOSUB 4060
3662 PRINT "VSWR(f)=";(1+M11)/(1-M11)
3664 PRINT "C(f)[dB]=";20/ LOG(10)*LOG(1/M14)
3666 GOSUB 4232
3668 GOTO 3802
3670 PRINT
3672 PRINT "Results, see Fig. 5.10"
3674 PRINT
3676 PRINT "1-input port"
3678 PRINT "2-coupled port"
3680 PRINT "4-isolated port"
3682 PRINT
3684 LET CC=1/ SQR (10^(CO/10))
3686 LET ZCE=ZO* SQR ((1+CC)/(1-CC))
3688 LET ZCO=ZO*ZO/ZCE
3690 PRINT "Zoe[ohm]=";ZCE
3692 PRINT "Zoo[ohm]=";ZCO
3694 PRINT "O(fo)[rad]= PI/2"
3696 PRINT "v(fo)[rad]=PI/2"
3698 PRINT "VSWR(f)=1"
3700 LET O= PI *F/(2*FO)
3702 LET CF=10/ LOG(10)*LOG((1+(1-CC*CC)/( TAN(O)* TAN(O)))/(CC
*CC))
3704 PRINT "C(f)[dB]=";CF
3706 GOSUB 4232
3708 GOTO 3974
3710 INPUT "Zo[ohm]=", ZO
3712 INPUT "C[dB]=",CO
3714 INPUT "fc[Hz]=",FC
3716 PRINT
3718 IF ZO <= 0 THEN  PRINT "Error:Zo <= 0": GOTO 3710
3720 IF CO <= 0 THEN  PRINT "Error:C <= 0": GOTO 3710
3722 IF FC <= 0 THEN  PRINT "Error:fc <= 0": GOTO 3710
3724 IF CO=8.34 THEN  LET Y=1: GOTO 3732
3726 IF CO=10 THEN  LET Y=2: GOTO 3732
3728 IF CO=20 THEN  LET Y=3: GOTO 3732
3730 PRINT "Please assume C equal to 8.34, 10 or 20 dB":PRINT:
GOTO 3710
```

```
3732 DIM Z(11)
3734 DIM K(3,8)
3736 FOR I=1 TO 3
3738 FOR J=1 TO 8
3740 READ K(I,J)
3742 NEXT J
3744 NEXT I
3746 DATA .6688,-.9751,-.6127,1.6194,-.4024,-.5828,.2938,.354
3748 DATA .5749,-.95,-.2495,1.2963,-.6825,-.0996,.1174,.358
3750 DATA .198,-.4121,.1856,.1304,-.1095,-.0044,.0141,.365
3752 PRINT
3754 PRINT "Results, see Fig. 5.11"
3756 PRINT
3758 PRINT "1-input port"
3760 PRINT "2-coupled port"
3762 PRINT "4-isolated port"
3764 GOSUB 4232
3766 FOR N=1 TO 11
3768 LET CC=0
3770 FOR M=1 TO 7
3772 LET CC=CC+K(Y,M)*((N-1)*.1)^(M-1)
3774 NEXT M
3776 PRINT "x/l=";.1*(N-1)
3778 LET Z(N)=ZO* SQR ((1+CC)/(1-CC))
3780 PRINT "Zoe[ohm]=";Z(N)
3782 PRINT "Zoo[ohm]=";ZO*ZO/Z(N)
3784 PRINT
3786 IF ( N MOD 5)=0 THEN PRINT "Press any key" ELSE GOTO 3790
3788 IF INKEY$="" THEN GOTO 3788 ELSE CLS
3790 NEXT N
3792 PRINT
3794 PRINT "l=";K(Y,8);"* wavelength at frequency fc"
3796 PRINT
3798 GOSUB 4232
3800 GOTO 3974
3802 REM ML or TL realization
3804 INPUT "Enter the name ( TL or ML ) of the realization tech
nique :",B$
3806 IF B$="TL" THEN  GOTO 3812
3808 IF B$="ML" THEN  GOTO 3892
3810 PRINT "Wrong command !"
3812 PRINT
3814 PRINT "Stripline realization"
3816 PRINT
3818 INPUT "er=",ER
3820 INPUT "ur=",UR
3822 INPUT "b[m]=",B
3824 INPUT "t[m]=",T
3826 PRINT
3828 IF A$="DC/I" THEN  GOTO 3832
3830 IF A$="DC/II" THEN  GOTO 3864
3832 PRINT "Geometrical dimensions in meters, see Fig. 5.8(b)"
3834 PRINT
3836 LET ZC=ZO
3838 GOSUB 8252
```

```
3840 PRINT "Wo=";W
3842 LET ZC=ZO/Y1
3844 GOSUB 8252
3846 PRINT "W1=";W
3848 PRINT "134=";75*1000000!/( SQR (ER*UR)*FO)
3850 PRINT "112=";225*1000000!/( SQR (ER*UR)*FO)
3852 LET ZC=ZO/Y2
3854 GOSUB 8252
3856 PRINT "W2=";W
3858 PRINT "113=";75*1000000!/( SQR (ER*UR)*FO)
3860 PRINT "124=";75*1000000!/( SQR (ER*UR)*FO)
3862 GOTO 4222
3864 PRINT "Geometrical dimensions in meters, see Fig. 5.9(b)"
3866 PRINT
3868 LET ZC=ZO
3870 GOSUB 8252
3872 PRINT "Wo=";W
3874 LET ZC=ZO/Y1
3876 GOSUB 8252
3878 PRINT "W1=";W
3880 PRINT "11=";75*1000000!/( SQR (ER*UR)*FO)
3882 LET ZC=ZO/Y2
3884 GOSUB 8252
3886 PRINT "W2=";W
3888 PRINT "12=";75*1000000!/( SQR (ER*UR)*FO)
3890 GOTO 4222
3892 PRINT
3894 PRINT "Microstrip line realization"
3896 PRINT
3898 INPUT "er=",ER
3900 PRINT "ur=1"
3902 INPUT "h[m]=",H
3904 INPUT "t[m]=",T
3906 PRINT
3908 LET F=FO
3910 IF A$="DC/I" THEN  GOTO 3914
3912 IF A$="DC/II" THEN  GOTO 3946
3914 PRINT "Geometrical dimensions in meters, see Fig. 5.8(b)"
3916 PRINT
3918 LET ZC=ZO
3920 GOSUB 8454
3922 PRINT "Wo=";(U-DTU)*H
3924 LET ZC=ZO/Y1
3926 GOSUB 8454
3928 PRINT "W1=";(U-DTU)*H
3930 PRINT "134=";75*1000000!/( SQR (EF)*FO)
3932 PRINT "112=";225*1000000!/( SQR (EF)*FO)
3934 LET ZC=ZO/Y2
3936 GOSUB 8454
3938 PRINT "W2=";(U-DTU)*H
3940 PRINT "113=";75*1000000!/( SQR (EF)*FO)
3942 PRINT "124=";75*1000000!/( SQR (EF)*FO)
3944 GOTO 4222
3946 PRINT "Geometrical dimensions in meters, see Fig. 5.9(b)"
3948 PRINT
```

```
3950 LET ZC=ZO
3952 GOSUB 8454
3954 PRINT "Wo=";(U-DTU)*H
3956 LET ZC=ZO/Y1
3958 GOSUB 8454
3960 PRINT "W1=";(U-DTU)*H
3962 PRINT "l1=";75*1000000!/( SQR (EF)*FO)
3964 LET ZC=ZO/Y2
3966 GOSUB 8454
3968 PRINT "W2=";(U-DTU)*H
3970 PRINT "l2=";75*1000000!/( SQR (EF)*FO)
3972 GOTO 4222
3974 PRINT
3976 INPUT "Enter the name ( CTL ) of the realization technique
:",B$
3978 IF B$="CTL" THEN  GOTO 3982
3980 PRINT "Wrong command !":GOTO 3974
3982 PRINT
3984 PRINT "Coupled striplines realization"
3986 PRINT
3988 INPUT "er=",ER
3990 INPUT "ur=",UR
3992 INPUT "b[m]=",B
3994 INPUT "t[m]=",T
3996 PRINT
3998 IF A$="DC/III" THEN  GOTO 4002
4000 IF A$="DC/IV" THEN  GOTO 4022
4002 PRINT "Geometrical dimensions in meters, see Fig. 5.10(b)"
4004 PRINT
4006 LET ZC=ZO
4008 GOSUB 8252
4010 PRINT "Wo=";W
4012 GOSUB 9360
4014 PRINT "W=";W
4016 PRINT "S=";S
4018 PRINT "lo=";75*1000000!/( SQR (ER*UR)*FO)
4020 GOTO 4222
4022 PRINT "Geometrical dimensions in meters, see Fig. 5.11(b)"
4024 PRINT
4026 LET ZC=ZO
4028 GOSUB 8252
4030 PRINT "Wo=";W
4032 PRINT
4034 FOR N=1 TO 11
4036 LET ZCE=Z(N)
4038 LET ZCO=ZO*ZO/ZCE
4040 GOSUB 9360
4042 PRINT "x/l=";.1*(N-1)
4044 PRINT "W=";W
4046 PRINT "S=";S
4048 PRINT
4050 IF ( N MOD 5 )=0 THEN PRINT"Press any key" ELSE GOTO 4054
4052 IF INKEY$="" THEN GOTO 4052 ELSE CLS
4054 NEXT N
4056 PRINT "l=";K(Y,3)*3*1E+08/( SQR (ER*UR)*FC)
```

```
4058 GOTO 4222
4060 REM S-parameters
4062 LET O= PI *F/(4*FO)
4064 LET A1=Y2*Y2-Y1*Y1* TAN (K*O)* TAN (O)
4066 LET A2=Y2*Y2-Y1*Y1/( TAN (K*O)* TAN (O))
4068 LET B1=Y1*Y2*( TAN (K*O)+ TAN (O))/ TAN (2*O)
4070 LET B2=Y1*Y2*(1/ TAN (K*O)+1/ TAN (O))/ TAN (2*O)
4072 LET C1=Y1*( TAN (K*O)+ TAN (O))
4074 LET C2=Y1*(1/ TAN (K*O)+1/ TAN (O))
4076 LET D1=Y1*( TAN (K*O)- TAN (O))
4078 LET D2=Y1*(1/ TAN (K*O)-1/ TAN (O))
4080 LET E=-2*Y2/ TAN (2*O)
4082 LET RN=1-A1-B1
4084 LET XN=-D1
4086 LET RD=1+A1+B1
4088 LET XD=C1+E
4090 LET R11E=(RN*RD+XN*XD)/(RD*RD+XD*XD)
4092 LET X11E=(XN*RD-XD*RN)/(RD*RD+XD*XD)
4094 LET XN=D1
4096 LET R22E=(RN*RD+XN*XD)/(RD*RD+XD*XD)
4098 LET X22E=(XN*RD-XD*RN)/(RD*RD+XD*XD)
4100 LET RN=0
4102 LET XN=-2*Y2/ SIN (2*O)
4104 LET R12E=XN*XD/(RD*RD+XD*XD)
4106 LET X12E=XN*RD/(RD*RD+XD*XD)
4108 LET RN=1-A2+B2
4110 LET XN=D2
4112 LET RD=1+A2-B2
4114 LET XD=E-C2
4116 LET R11O=(RN*RD+XN*XD)/(RD*RD+XD*XD)
4118 LET X11O=(XN*RD-XD*RN)/(RD*RD+XD*XD)
4120 LET XN=-D2
4122 LET R22O=(RN*RD+XN*XD)/(RD*RD+XD*XD)
4124 LET X22O=(XN*RD-XD*RN)/(RD*RD+XD*XD)
4126 LET RN=0
4128 LET XN=-2*Y2/ SIN (2*O)
4130 LET R12O=XN*XD/(RD*RD+XD*XD)
4132 LET X12O=XN*RD/(RD*RD+XD*XD)
4134 LET R=(R11E+R11O)/2
4136 LET X=(X11E+X11O)/2
4138 GOSUB 4196
4140 LET M11=M
4142 LET V11=V
4144 LET R=(R22E+R22O)/2
4146 LET X=(X22E+X22O)/2
4148 GOSUB 4196
4150 LET M33=M
4152 LET V33=V
4154 LET R=(R11E-R11O)/2
4156 LET X=(X11E-X11O)/2
4158 GOSUB 4196
4160 LET M12=M
4162 LET V12=V
4164 LET R=(R22E-R22O)/2
4166 LET X=(X22E-X22O)/2
```

```
4168 GOSUB 4196
4170 LET M34=M
4172 LET V34=V
4174 LET R=(R12E+R12O)/2
4176 LET X=(X12E+X12O)/2
4178 GOSUB 4196
4180 LET M13=M
4182 LET V13=V
4184 LET R=(R12E-R12O)/2
4186 LET X=(X12E-X12O)/2
4188 GOSUB 4196
4190 LET M14=M
4192 LET V14=V
4194 RETURN
4196 REM Subroutine 4196
4198 LET M= SQR (R*R+X*X)
4200 IF M=0 THEN  LET V=0: RETURN
4202 IF (X/M) >= .999999 THEN  LET V= PI /2: RETURN
4204 IF (X/M) <= -.999999 THEN  LET V=- PI /2: RETURN
4206 IF  ABS (R) <= 1E-09 THEN  GOTO 4214
4208 IF R<0 THEN  GOTO 4212
4210 LET V= ATN (X/R): RETURN
4212 LET V= PI + ATN (X/R): RETURN
4214 IF X=0 AND R>0 THEN  LET V=0: RETURN
4216 IF X=0 AND R<0 THEN  LET V= PI : RETURN
4218 IF X>0 THEN  LET V= PI /2- ATN (R/X): RETURN
4220 LET V=- PI /2+ ATN (R/X): RETURN
4222 REM Chain subroutine
4224 PRINT
4226 INPUT"Enter 'CONT':",N$
4228 IF N$="CONT" THEN CLS:CHAIN "MENU",10
4230 GOTO 4224
4232 REM Control subroutine
4234 PRINT
4236 INPUT "Enter 'CONT':",N$
4238 IF N$="CONT" THEN CLS:RETURN
4240 GOTO 4234
```

5.5 THREE-PORT HYBRID POWER DIVIDERS

The computer program HPD, presented in this section, is used for designing the three-port TEM-mode hybrids which are useful both as power dividers and power combiners (see e.g., [14–16]). The electrical scheme and structural outline of the one-section hybrid are shown in Figure 5.14. Similarly, Figure 5.15 presents the general electrical scheme and the microstrip line layout of the three-port hybrids, which consist of many quarter-wavelength sections. In general, the design of these circuits uses the following input data:

Z_0 = the terminating real impedances in ohms,
f_0 = the center frequency of the required passband in Hz,

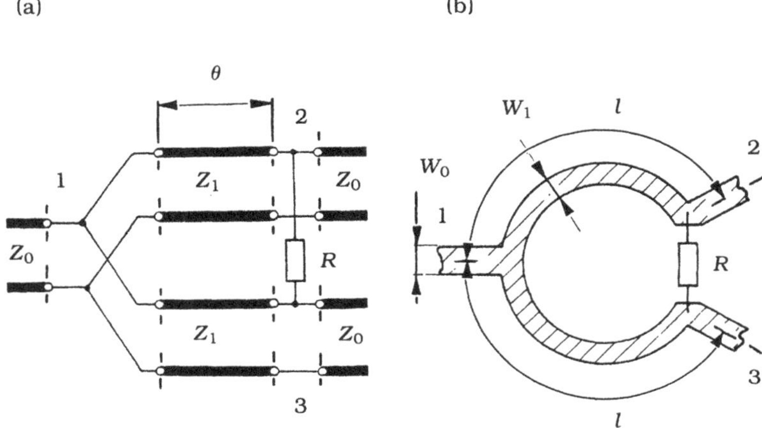

Figure 5.14 The three-port hybrid power divider: (a) electrical scheme and (b) microstrip line configuration.

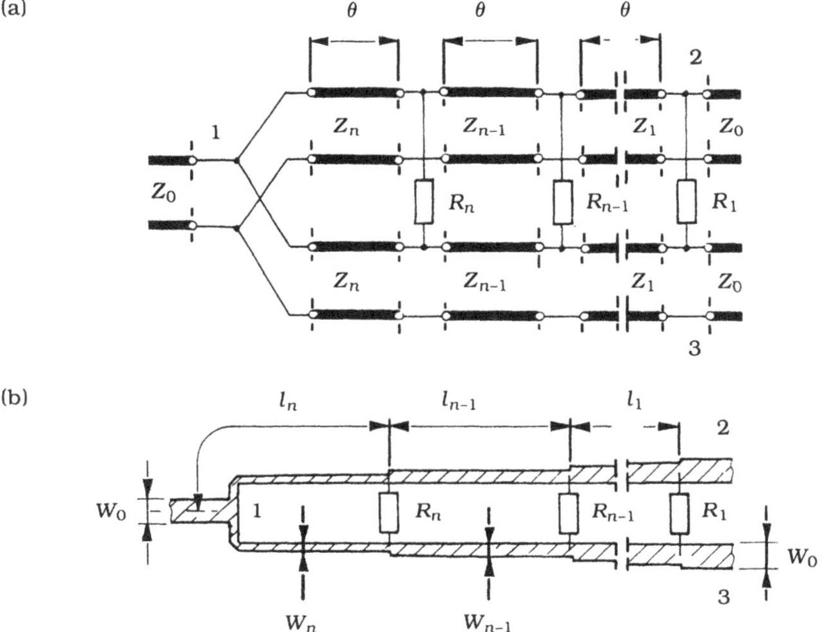

Figure 5.15 The multisection three-port hybrid power divider: (a) electrical scheme and (b) microstrip line configuration.

w = the fractional bandwidth ($0 < w \leqslant 1.2$)
VSWR = the maximum value of the voltage standing wave ratio over the required passband,
I = the minimum value of the isolation between output ports over the required passband in dB.

The above input data are verified at the beginning of the design process. If they are acceptable for the design, then the one-section hybrid is analyzed first. Its frequency responses VSWR(θ) and $I(\theta)$ are calculated from the following formulas:

$$\text{VSWR}(\theta) = \frac{1 + |S_{11}|}{1 - |S_{11}|} = \frac{\sqrt{8t^2 + 9} + 1}{\sqrt{8t^2 + 9} - 1}$$
$$I(\theta) = 20 \log \left| \frac{1}{S_{23}} \right| = 10 \log \left[\frac{64t^4 + 80t^2 + 9}{4(2t^2 + 1)} \right], \text{ dB} \tag{5.43}$$

where $t = \tan(\theta)$, and θ is the electrical length of the sections. The given values of VSWR and I are compared with the corresponding values of VSWR(θ) and $I(\theta)$ computed at point $\theta_r = \pi(2 - w)/4$ rad. If VSWR(θ_r) $\leqslant$ VSWR and $I(\theta_r) \geqslant I$, then the analyzed divider satisfies the assumed requirements. For this divider (see Figure 5.14), $Z_1 = \sqrt{2}\, Z_0$, $R = 2Z_0$ and $\theta(f_0) = \pi/2$ radians. In other cases, the multisection hybrids are analyzed with respect to Cohn's design data given in Table 5.4 [16]. Consequently, we choose the divider with minimum number of sections $n \leqslant 4$, which satisfies the assumed electrical requirements. Next the characteristic impedances $Z_i = z_i Z_0$, where $i = 1, 2, \ldots, n$, of the individual sections are calculated on the basis of suitable data z_i given in Table 5.4. Finally, the interconnecting resistances $R_i = r_i Z_0$, where $i = 1, 2, \ldots, n$, are calculated in a similar manner.

Table 5.4
Performance Limits and Normalized Parameters of Three-Port Hybrid Power Dividers

n	2	2	3	3	4
w	0.400	0.666	0.666	1.000	1.200
VSWR_{max}	1.036	1.106	1.029	1.105	1.100
I_{min}, dB	36.6	27.3	38.7	27.9	26.8
z_1	1.1998	1.2197	1.1124	1.1497	1.1157
z_2	1.6670	1.6398	1.4142	1.4142	1.2957
z_3	—	—	1.7979	1.7396	1.5435
z_4	—	—	—	—	1.7926
r_1	5.3163	4.8204	10.0000	8.0000	9.6432
r_2	1.8643	1.9602	3.7460	4.2292	5.8326
r_3	—	—	1.9048	2.1436	3.4524
r_4	—	—	—	—	2.0633

The program HPD (see flow diagram shown in Figure 5.16) also makes possible the evaluation of the geometrical dimensions of the divider if the microstrip line technique was used for the design. For this purpose, we must introduce into the computer memory suitable structural data. After this, program ML continues the calculations. This program is described in Section 6.5, and is called out $(n + 1)$ times as a subroutine.

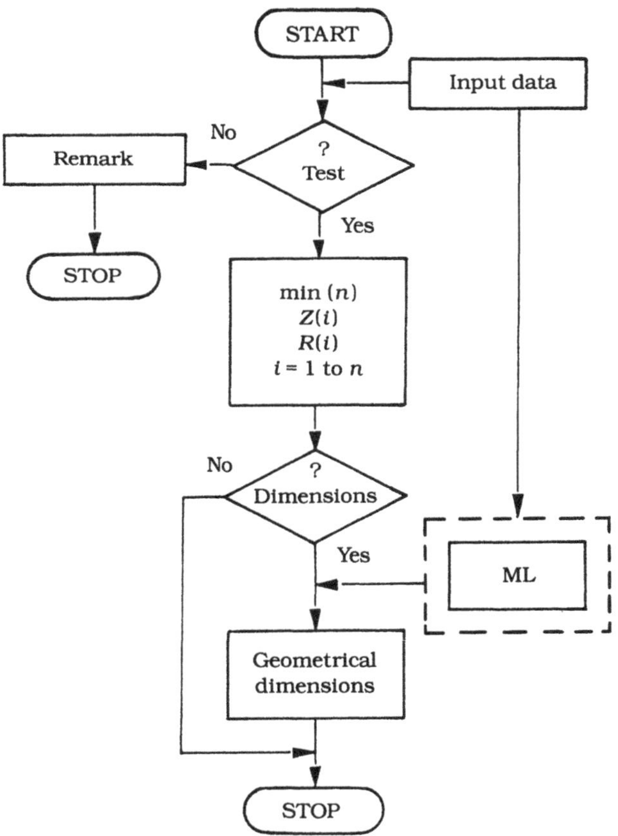

Figure 5.16 The flow diagram of the HPD computer program.

Check Calculations

Table 5.5 contains some typical results of calculations obtained by using the program HPD. These results have been obtained for the following input data: $Z_0 = 50\ \Omega$, $f_0 = 10^9$ Hz, $w = 0.9$, VSWR $= 1.2$ and $I = 26$ dB. We have assumed that

Table 5.5
Electrical Parameters and Geometrical Dimensions of the Three-Section Three-Port Hybrid Power Divider

	$n = 3,\ W_0 = 2.90\ 10^{-3}\ m$			
i	Z_i, ohm	R_i, ohm	W_i, m	l_i, m
1	57.48	400.00	$2.27\ 10^{-3}$	$42.07\ 10^{-3}$
2	70.71	211.46	$1.52\ 10^{-3}$	$42.70\ 10^{-3}$
3	86.98	107.18	$0.94\ 10^{-3}$	$43.31\ 10^{-3}$

the divider is constructed of microstrip line segments by using a dielectric substrate with $\epsilon_r = 4.24$, $\mu_r = 1$, $h = 1.5 \cdot 10^{-3}$ m and $t = 3.5 \cdot 10^{-5}$ m (see Figures 6.4 and 5.15).

Listing of Program HPD

```
4400 REM HPD
4402 CLS
4404 CHAIN MERGE "ML", 4406
4406 PRINT "HYBRID POWER DIVIDERS"
4408 DIM Z(4)
4410 DIM R(4)
4412 PRINT
4414 PRINT "Data:"
4416 PRINT
4418 INPUT "Zo[ohm]=",ZO
4420 INPUT "fo[Hz]=",FO
4422 INPUT "w=",W
4424 INPUT "VSWR=",QM
4426 INPUT "IdB]=",IM
4428 IF ZO <= 0 THEN  PRINT "Error:Zo <= 0": GOTO 4416
4430 IF W <= 0 THEN  PRINT "Error:w <= 0": GOTO 4416
4432 IF W >= 2 THEN  PRINT "Error:w >= 2": GOTO 4416
4434 IF QM <= 0 THEN  PRINT "Error:I <= 0": GOTO 4416
4436 LET PI=3.141593
4438 PRINT
4440 PRINT "Please wait !"
4442 PRINT
4444 IF W>1.2 OR IM>40 THEN  PRINT "Please assume w<1.2 and I<4
0 dB": PRINT : GOTO 4412
4446 LET T= TAN ( PI /2*(1-W/2))
4448 LET V=T*T
4450 LET Q=( SQR (9+8*V)+1)/( SQR (9+8*V)-1)
4452 LET I=20/ LOG(10)* LOG( SQR (64*V*V+90*V+9)/(2* SQR (2*V+1
)))
4454 IF Q <= QM AND I >= IM THEN  LET N=1: GOTO 4646
4456 IF W <= .666 THEN  GOTO 4462
4458 IF W <= 1 THEN  GOTO 4518
4460 GOTO 4590
```

```
4462 LET N=2
4464 IF W <= .4 AND QM>1.036 THEN  GOTO 4492
4466 IF QM<1.106 OR I >= 27.3 THEN  GOTO 4518
4468 PRINT "Results, see Fig. 5.15"
4470 PRINT
4472 PRINT "n=2"
4474 PRINT
4476 LET Z(1)=ZO*1.2197
4478 PRINT "Z(1)[ohm]=";Z(1)
4480 LET Z(2)=ZO*1.6398
4482 PRINT "Z(2)[ohm]=";Z(2)
4484 PRINT
4486 PRINT "R(1)[ohm]=";ZO*4.8204
4488 PRINT "R(2)[ohm]=";ZO*1.9602
4490 GOTO 4662
4492 IF IM>36.8 THEN  GOTO 4518
4494 PRINT "Results, see Fig. 5.15"
4496 PRINT
4498 PRINT "n=2"
4500 PRINT
4502 LET Z(1)=ZO*1.1998
4504 PRINT "Z(1)[ohm]=";Z(1)
4506 LET Z(2)=ZO*1.667
4508 PRINT "Z(2)[ohm]=";Z(2)
4510 PRINT
4512 PRINT "R(1)[ohm]=";ZO*5.3163
4514 PRINT "R(2)[ohm]=";ZO*1.8643
4516 GOTO 4662
4518 LET N=3
4520 IF W <= .666 AND QM>1.029 THEN  GOTO 4556
4522 IF QM<1.105 OR IM>27.9 THEN  PRINT "Please assume VSWR>1.1
05 or I<27.9 dB": PRINT : GOTO 4412
4524 LET Z(1)=ZO*1.1497
4526 PRINT
4528 PRINT "Results, see Fig. 5.15"
4530 PRINT
4532 PRINT "n=3"
4534 PRINT
4536 PRINT "Z(1)[ohm]=";Z(1)
4538 LET Z(2)=ZO*1.4142
4540 PRINT "Z(2)[ohm]=";Z(2)
4542 LET Z(3)=ZO*1.7396
4544 PRINT "Z(3)[ohm]=";Z(3)
4546 PRINT
4548 PRINT "R(1)[ohm]=";ZO*8
4550 PRINT "R(2)[ohm]=";ZO*4.2292
4552 PRINT "R(3)[ohm]=";ZO*2.1436
4554 GOTO 4662
4556 IF IM>38.7 THEN  PRINT "Please assume I<38.7 dB": PRINT :
GOTO 4412
4558 PRINT
4560 PRINT "Results, see Fig. 5.15"
4562 PRINT
4564 PRINT "n=3"
4566 PRINT
```

```
4568 LET Z(1)=ZO*1.1124
4570 PRINT "Z(1)[ohm]=";Z(1)
4572 LET Z(2)=ZO*1.4142
4574 PRINT "Z(2)[ohm]=";Z(2)
4576 LET Z(3)=ZO*1.7979
4578 PRINT "Z(3)[ohm]=";Z(3)
4580 PRINT
4582 PRINT "R(1)[ohm]=";ZO*10
4584 PRINT "R(2)[ohm]=";ZO*3.746
4586 PRINT "R(3)[ohm]=";ZO*1.9048
4588 GOTO 4662
4590 LET N=4
4592 IF QM>1.1 AND IM <= 26.8 THEN  GOTO 4610
4594 PRINT "Suggestion"
4596 PRINT
4598 PRINT "n=4"
4600 PRINT
4602 PRINT "VSWR <= 1.1"
4604 PRINT "I <= 26.8"
4606 PRINT
4608 GOTO 4618
4610 PRINT "Results, see Fig. 5.15"
4612 PRINT
4614 PRINT "n=4"
4616 PRINT
4618 LET Z(1)=ZO*1.1157
4620 PRINT "Z(1)[ohm]=";Z(1)
4622 LET Z(2)=ZO*1.2957
4624 PRINT "Z(2)[ohm]=";Z(2)
4626 LET Z(3)=ZO*1.5435
4628 PRINT "Z(3)[ohm]=";Z(3)
4630 LET Z(4)=ZO*1.7926
4632 PRINT "Z(4)[ohm]=";Z(4)
4634 PRINT
4636 PRINT "R(1)[ohm]=";ZO*9.6432
4638 PRINT "R(2)[ohm]=";ZO*5.8326
4640 PRINT "R(3)[ohm]=";ZO*3.4524
4642 PRINT "R(4)[ohm]=";ZO*2.0633
4644 GOTO 4662
4646 PRINT "Results, see Fig. 5.14"
4648 PRINT
4650 PRINT "n=1"
4652 PRINT
4654 LET Z(1)=ZO* SQR(2)
4656 PRINT "Z(1)[ohm]=";Z(1)
4658 PRINT
4660 PRINT "R(1)[ohm]=";ZO*2
4662 LET NN=N
4664 GOSUB 4716
4666 PRINT
4668 PRINT "Microstrip line realization"
4670 PRINT
4672 INPUT "er=",ER
4674 INPUT "h[m]=",H
```

```
4676 INPUT "t[m]=",T
4678 PRINT
4680 PRINT"Geometrical dimensions in meters, see Figs. 5.14 and
 5.15"
4682 LET F=FO
4684 FOR J=1 TO NN
4686 LET ZC=Z(J)
4688 GOSUB 8454
4690 PRINT
4692 PRINT "W(";J;")[m]=";(U-DTU)*H
4694 PRINT "l(";J;")[m]=";75*10^6/(FO* SQR ( EF))
4696 NEXT J
4698 PRINT
4700 LET ZC=ZO
4702 GOSUB 8454
4704 PRINT"Wo[M]=";(U-DTU)*H
4706 REM Chain subroutine
4708 PRINT
4710 INPUT"Enter 'CONT':",N$
4712 IF N$="CONT" THEN CLS:CHAIN "MENU",10
4714 GOTO 4708
4716 REM Control subroutine
4718 PRINT
4720 INPUT "Enter 'CONT':",N$
4722 IF N$="CONT" THEN CLS:RETURN
4724 GOTO 4718
```

5.6 LUMPED AND DISTRIBUTED LOW-PASS FILTERS

The computer program LPF, presented in this section, is meant for the design of the Chebyschev low-pass filters the insertion loss function and electrical diagram of which are shown in Figures 5.17 and 5.18, respectively.

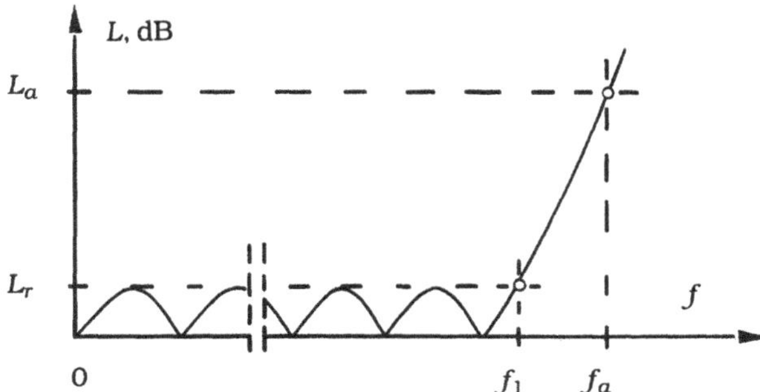

Figure 5.17 The typical insertion loss function for the Chebyschev low-pass filters.

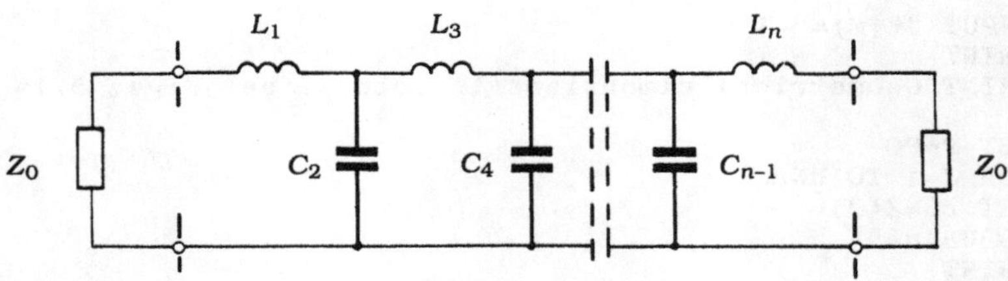

Figure 5.18 The lumped element electrical scheme of a low-pass filter.

At the beginning of the design process, we must introduce into the computer memory the following data:

Z_0 = the terminating resistances in ohms,
f_1 = the cut-off frequency in Hz,
L_r = the maximum tolerable attenuation over the passband in dB,
f_a = the reference frequency ($f_a > f_1$) in Hz,
L_a = the attenuation inserted by the filter at frequency f_a, expressed in dB.

In the first design step, the computer evaluates the minimum odd number n of reactances of the low-pass prototype filter from the condition:

$$n \geqslant \frac{\cosh^{-1}\sqrt{(L'_a - 1)/(L'_r - 1)}}{\cosh^{-1}(f_a/f_1)} \tag{5.44}$$

where $L'_a = 10^{(L_a/10)}$ and $L'_r = 10^{(L_r/10)}$. Here, the mathematical identity:

$$\cosh^{-1}(x) = \ln(x + \sqrt{x^2 - 1}) \quad \text{for } x > 1$$

enables rapid evaluation of (5.44). When the minimum odd number n is known, values of the filter inductances and capacitances are calculated as:

$$\begin{aligned} L_i &= Z_0 g_i/\omega_1 \quad \text{for } i = 1, 3, \ldots, n \\ C_j &= g_j/(Z_0\omega_1) \quad \text{for } j = 2, 4, \ldots, n-1 \end{aligned} \tag{5.45}$$

where $\omega_1 = 2\pi f_1$ and $g_1, g_2, g_3, \ldots, g_n$ are the suitable elements of the Chebyschev prototype filter. According to [5], these elements can be calculated from the following iterative formulas:

$$
\begin{aligned}
g_0 &= 1 \\
g_1 &= 2a_1/y \\
g_k &= \frac{4a_{k-1}a_k}{b_{k-1}g_{k-1}}, \qquad \text{for } k = 2, 3, \ldots, n \\
g_{n+1} &= \begin{cases} 1, & \text{for odd } n \\ \coth^2(x/4), & \text{for even } n \end{cases}
\end{aligned} \tag{5.46}
$$

where

$$
\begin{aligned}
x &= \ln[\coth(L_r/17.37)] \\
y &= \sinh[x/(2\text{n})] \\
a_k &= \sin[\pi(2k - 1)/(2n)] \\
b_k &= y^2 + \sin^2(k\pi/n)
\end{aligned}
$$

Note that if n is odd, the filter components will display symmetry. For even n, the component values generated will lead to an asymmetric design. This may be useful, especially when the terminating resistances are unequal. This completes the design of the filter composed of lumped inductances and capacitances.

At microwave frequencies, however, the lumped filter components tend to become small, and therefore may be difficult to realize. In addition, as the frequency is increased, the lumped inductances and capacitances depart from their ideal characteristics due to radiation, loss mechanisms, and propagation effects. For these reasons, at microwave frequencies (when the length of lines becomes close to that of the wavelength, see [5,20]), the high-impedance and low-impedance segments of the transmission lines can be used instead of the corresponding lumped inductors and capacitors. The lengths and characteristic impedances of these line segments should be carefully selected to simulate as closely as possible the behavior of the lumped elements. Figure 5.19 shows how the lumped reactances can be realized in the form of the TEM transmission line segment. Outlines of typical filters constructed in this way are shown in Figure 5.20. The geometrical dimensions of the filter constructed of segments of coaxial (ECL) and microstrip (ML) lines can be evaluated automatically by using the program LPF again. After introducing the command CONTINUE, the computer asks for the line name and suitable additional data. When the coaxial line technique is chosen, the computer asks for:

Z_l = minimum value of the characteristic impedance of the line segments in ohms,

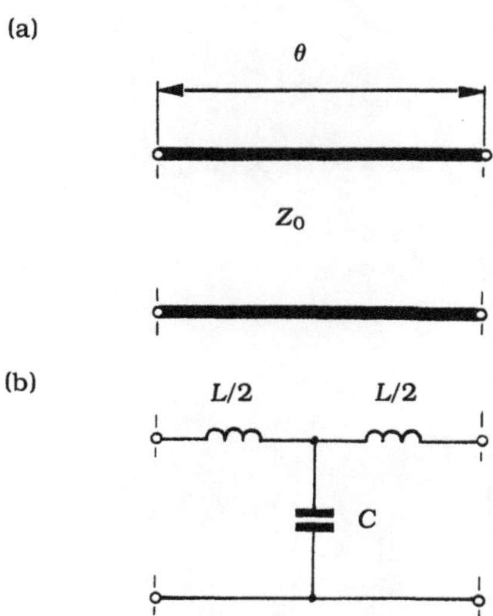

Figure 5.19 The short segment of the TEM transmission line and its equivalent electrical scheme, where $C = \sin(\theta)/(\omega Z_0)$ and $L/2 = Z_0 \tan(\theta/2)/\omega$.

Z_h = maximum value of the characteristic impedance of the line segments in ohms,
D = diameter of the outer conductor of the coaxial line in m,
ϵ_r = relative permittivity of the dielectric substrate (see Figure 5.20(a)).

Similarly, for the microstrip line technique, the following data are needed:

Z_l = minimum value of the characteristic impedance of the line segments in ohms,
Z_h = maximum value of the characteristic impedance of the line segments in ohms,
C_f = equivalent capacitance of the $Z_l - Z_h$ impedance step discontinuity in F,
C_{f0} = equivalent capacitance of the $Z_0 - Z_h$ impedance step discontinuity in F,
ϵ_r = relative permittivity of the line dielectric substrate (see Figure 5.20(b)),
h = dielectric substrate thickness in m,
t = strip thickness in m (see Figure 6.4).

(a)

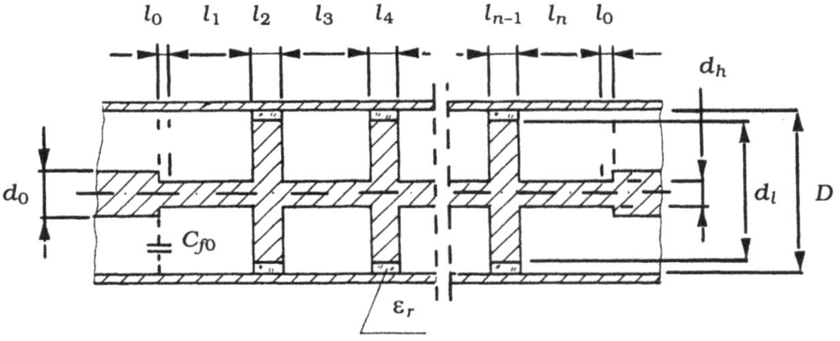

(b)

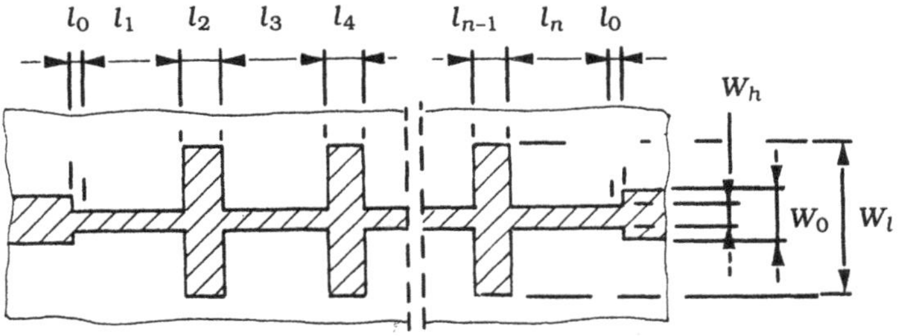

Figure 5.20 The microwave low-pass filter: (a) coaxial line longitudinal section and (b) microstrip line configuration.

The geometrical lengths $l_1, l_3, \ldots, l_n$ of the high-impedance line segments, which serve as inductors, are evaluated by solving the following system of equations:

$$
\begin{aligned}
\omega_1 L_1 &= Z_h \sin\left(\frac{\omega_1 l_1}{v_h}\right) + \frac{Z_1 l_2 \omega_1}{2v_l} \\
\omega_1 L_3 &= Z_h \sin\left(\frac{\omega_1 l_3}{v_h}\right) + \frac{Z_l l_2 \omega_1}{2v_l} + \frac{Z_1 l_4 \omega_1}{2v_l}
\end{aligned}
\tag{5.47}
$$

et cetera, where v_h and v_l are the velocities of propagation of a wave along the high- and low-impedance line segments, respectively.

Lengths $l_2, l_4, \ldots, l_{n-1}$ of the low-impedance line segments, which serve as capacitors, are calculated in a similar way (i.e., by solving the following system of equations):

$$\omega_1 C_2 = \frac{Y_l l_2 \omega_1}{v_l} + 2C_f \omega_1 + \frac{Y_h l_1 \omega_1}{2v_h} + \frac{Y_h l_3 \omega_1}{2v_h}$$
$$\omega_1 C_4 = \frac{Y_l l_4 \omega_1}{v_l} + 2C_f \omega_1 + \frac{Y_h l_3 \omega_1}{2v_h} + \frac{Y_h l_5 \omega_1}{2v_h} \tag{5.48}$$

et cetera.

In computer program LPF, (5.47) and (5.48) are solved iteratively (three iterations are made) for lengths $l_1, l_2, \ldots, l_n$ in the manner described in the monograph [5, pp. 369–371]. The discontinuity capacitances C_{f0} and $Y_h l_1/(2v_h)$ at the junction between the Z_0 impedance line and Z_h impedance line comprising the first inductive element (see Figures 5.19 and 5.20) are compensated for by increasing the length l_1 of the Z_h impedance segments by a small value l_0 calculated as follows [5]:

$$l_0 = Z_0^2[C_{f0}v_h/Z_h + l_1/(2Z_h^2)] \tag{5.49}$$

Finally, the corrected geometrical lengths of the first and last filter sections are $l_{1c} = l_{nc} = l_1 + l_0$.

The flow diagram of computer program LPF is shown in Figure 5.21.

Check Calculations

Computer program LPF has been tested for the following data: $Z_0 = 50\ \Omega$, $f_1 = 10^9$ Hz, $L_r = 0.2$ dB, $f_a = 1.42\ 10^9$ Hz and $L_a = 30$ dB. The values of the filter inductances and capacitances obtained for these data are given in the second and third columns of Table 5.6.

Table 5.6
Electrical Parameters and Geometrical Dimensions of the Coaxial Line Low-Pass Filter

	d_0 = 4.34 10^{-3} m, l_0 = 1.23 10^{-3} m (See Figures 5.18 and 5.20)				
i	*L_i, H*	*C_i, F*	*l_i, m*	*d_l, m*	*d_h, m*
1	10.92 10^{-9}	—	22.25 10^{-3}	—	0.82 10^{-3}
2	—	4.39 10^{-12}	7.74 10^{-3}	7.88 10^{-3}	—
3	18.11 10^{-9}	—	39.94 10^{-3}	—	0.82 10^{-3}
4	—	4.77 10^{-12}	8.14 10^{-3}	7.88 10^{-3}	—
5	18.11 10^{-9}	—	39.94 10^{-3}	—	0.82 10^{-3}
6	—	4.39 10^{-12}	7.74 10^{-3}	7.88 10^{-3}	—
7	10.92 10^{-9}	—	22.25 10^{-3}	—	0.82 10^{-3}

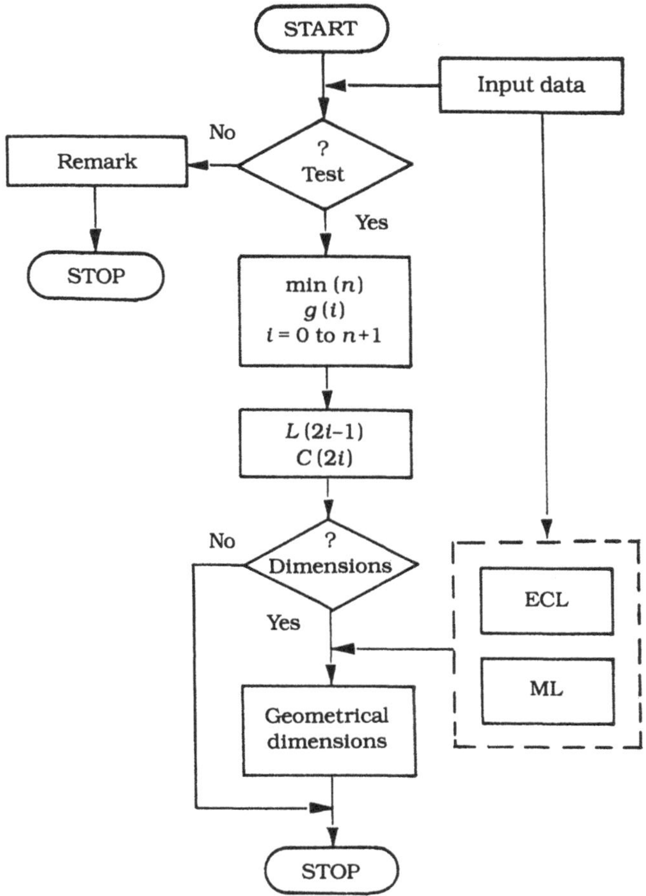

Figure 5.21 The flow diagram of the LPF computer program.

For the coaxial line technique design, the additional input data are: CL, $Z_l = 10\ \Omega$, $Z_h = 150\ \Omega$, $D = 10^{-2}$ m and $\epsilon_r = 2.05$. In this case, we obtain as output results the main geometrical dimensions of the filter whose longitudinal section is shown in Figure 5.20(a). The values of these dimensions are presented in the remaining columns of Table 5.6. Here, we note that the discontinuity capacitances C_f and C_{f0} needed in the design process have been calculated by using the following approximate formula:

$$C = 10^{-12}\pi D\left(13 + \frac{D}{5d_1}\right)\left(\frac{d_2 - d_1}{D - d_1}\right)^{2.65}, \text{F} \qquad (5.50)$$

where D, d_1 and d_2 are diameters of the coaxial line step, as shown in Figure 5.22. These diameters should be expressed in meters. Experiments have shown that (5.50) is practically adequate if $1 < D/d_1 \leqslant 15$.

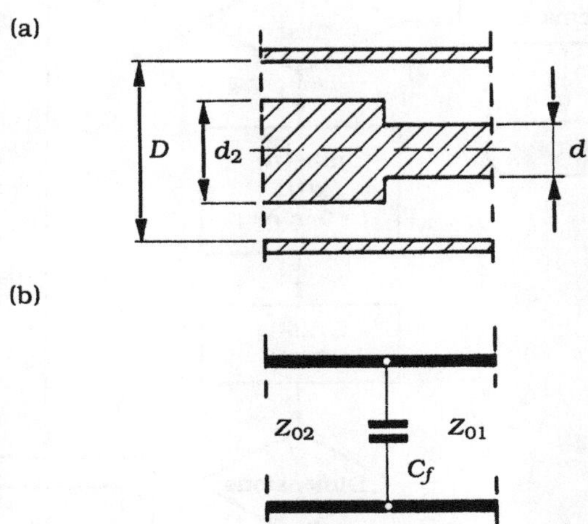

Figure 5.22 The coaxial line step discontinuity: (a) longitudinal section and (b) electrical scheme.

Listing of Program LPF

```
4800 REM LPF
4802 CLS
4804 CHAIN MERGE "ML",4806
4806 PRINT "LOW PASS FILTERS"
4808 PRINT
4810 PRINT "Data:"
4812 PRINT
4814 INPUT "Zo[ohm]=",ZO
4816 INPUT "f1[Hz]=",F1
4818 INPUT "Lr[dB]=",LR
4820 INPUT "fa[Hz]=",FA
4822 INPUT "La[dB]=",LA
4824 PRINT
4826 LET PI=3.141593
4828 IF ZO <= 0 THEN  PRINT "Error:Zo <= 0": GOTO 4812
4830 IF F1 <= 0 THEN  PRINT "Error:f1 <= 0": GOTO 4812
4832 IF LR <= 0 THEN  PRINT "Error:Lr <= 0": GOTO 4812
4834 IF FA <= F1 THEN  PRINT "Error:fa <= f1": GOTO 4812
4836 IF LA <= LR THEN  PRINT "Error:La <= Lr": GOTO 4812
4838 LET X= SQR ((10^(LA/10)-1)/(10^(LR/10)-1))
4840 LET A= LOG(X+ SQR (X*X-1))
```

```
4842 LET X=FA/F1
4844 LET B= LOG(X+ SQR (X*X-1))
4846 LET N= INT (.99+A/B)
4848 IF (N/2- INT (N/2+.001))>.03 THEN  GOTO 4852
4850 LET N=N+1
4852 DIM A(N)
4854 DIM B(N)
4856 DIM G(N+1)
4858 DIM L(N)
4860 DIM C(N)
4862 LET X= EXP (LR/17.37)
4864 LET C= LOG((X+1/X)/(X-1/X))
4866 LET X= EXP (C/(2*N))
4868 LET Y=(X-1/X)/2
4870 FOR I=1 TO N
4872 LET A(I)= SIN ((2*I-1)* PI /(2*N))
4874 LET B(I)=Y*Y+ SIN (I* PI /N)* SIN (I* PI /N)
4876 NEXT I
4878 LET G(1)=2*A(1)/Y
4880 FOR I=2 TO N
4882 LET G(I)=4*A(I-1)*A(I)/(B(I-1)*G(I-1))
4884 NEXT I
4886 LET G(N+1)=1
4888 REM Calculation of L(2*i-1) and C(2*i)
4890 FOR I=1 TO  INT (N/2)
4892 LET L(2*I-1)=ZO*G(2*I-1)/(2* PI *F1)
4894 LET C(2*I)=G(2*I)/(ZO*2* PI *F1)
4896 NEXT I
4898 LET L(N)=ZO*G(N)/(2* PI *F1)
4900 PRINT "Electrical parameters, see Fig. 5.18"
4902 PRINT
4904 PRINT "n=";N
4906 PRINT
4908 FOR I=1 TO  INT (N/2)
4910 PRINT "L(";2*I-1;")[H]=";L(2*I-1)
4912 PRINT "C(";2*I;")[F]=";C(2*I)
4914 NEXT I
4916 PRINT "L(";N;")[H]=";L(N)
4918 GOSUB 5180
4920 DIM X(N)
4922 FOR I=1 TO  INT (N/2)
4924 LET X(2*I-1)=G(2*I-1)*ZO
4926 LET X(2*I)=G(2*I)/ZO
4928 NEXT I
4930 LET X(N)=G(N)*ZO
4932 LET XH=X(1)
4934 FOR I=2 TO N
4936 IF XH<X(I) THEN  LET XH=X(I)
4938 NEXT I
4940 PRINT
4942 INPUT "Enter the name ( CL or ML ) of the realization tech
nique:",D$
4944 PRINT
4946 LET NN=N
4948 LET VO=299792500#
```

```
4950 IF D$="CL" THEN  GOTO 4958
4952 IF D$="ML" THEN  GOTO 5026
4954 PRINT "Wrong command !"
4956 GOTO 4940
4958 REM Coaxial line realization
4960 INPUT "D[m]=",D
4962 INPUT "er=",ER
4964 INPUT "Zl[ohm]=",ZL
4966 INPUT "Zh[ohm]=",ZH
4968 PRINT
4970 IF ZH>XH THEN  GOTO 4978
4972 PRINT "Please assume Zh[ohm]>";XH
4974 PRINT
4976 GOTO 4966
4978 IF ZL<ZO THEN  GOTO 4986
4980 PRINT "Please assume Zl[ohm]<";ZO
4982 PRINT
4984 GOTO 4964
4986 LET DO=D/ EXP (ZO/60)
4988 LET DH=D/ EXP (ZH/60)
4990 LET VH=VO
4992 LET DL=D/ EXP (ZL* SQR (ER)/60)
4994 LET VL=VO/ SQR (ER)
4996 LET CFO= PI *D*1E-12/(10^12)*(13+.2*D/DH)*((DO-DH)/(D-DH))
^2.65
4998 LET CF= PI *D*1E-12/(10^12)*(13+.2*D/DH)*((DL-DH)/(D-DH))^
2.65
5000 GOSUB 5128
5002 PRINT "Geometrical dimensions in meters Fig. 5.20(a) "
5004 PRINT
5006 PRINT "do=";DO
5008 PRINT "dh=";DH
5010 PRINT "dl=";DL
5012 PRINT "lo=";ZO*ZO*(CFO*VH/ZH+L(1)/(2*ZH*ZH))
5014 PRINT
5016 FOR I=1 TO NN
5018 PRINT "l(";I;")=";L(I)
5020 NEXT I
5022 PRINT
5024 GOTO 5170
5026 REM Microstrip line realization
5028 INPUT "er=",ER
5030 INPUT "h[m]=",H
5032 INPUT "t[m]=",T
5034 INPUT "Zl[ohm]=",ZL
5036 INPUT "Zh[ohm]=",ZH
5038 LET F=F1
5040 PRINT
5042 IF ZH>XH THEN  GOTO 5050
5044 PRINT "Please assume Zh[ohm]>";XH
5046 PRINT
5048 GOTO 5036
5050 IF ZL<ZO THEN  GOTO 5058
5052 PRINT "Please assume Zl[ohm]<";ZO
5054 PRINT
```

```
5056 GOTO 5034
5058 LET ZC=ZO
5060 PRINT "Please wait !"
5062 PRINT
5064 GOSUB 8454
5066 LET WO=(U-DTU)*H
5068 LET ZC=ZH
5070 GOSUB 8454
5072 LET WH=(U-DTU)*H
5074 LET VH=VO/ SQR (EF)
5076 LET ZC=ZL
5078 GOSUB 8454
5080 LET WL=(U-DTU)*H
5082 LET VL=VO/ SQR (EF)
5084 PRINT "Enter the value of discontinuity capacitace Cf for
Wl/Wh=";WL/WH
5086 PRINT
5088 INPUT "Cf[F]=",CF
5090 PRINT
5092 PRINT "Enter the value of discontinuity capacitace Cfo for
 Wo/Wh=";WO/WH
5094 PRINT
5096 INPUT "Cfo[F]=",CFO
5098 GOSUB 5128
5100 PRINT
5102 PRINT "Geometrical dimensions in meters, see Fig. 5.20(b)"
5104 PRINT
5106 PRINT "Wo=";WO
5108 PRINT "Wh=";WH
5110 PRINT "Wl=";WL
5112 PRINT
5114 FOR I=1 TO NN
5116 PRINT "l(";I;")=";L(I)
5118 NEXT I
5120 PRINT
5122 PRINT "lo=";ZO*ZO*(CFO*VH/ZH+L(1)/(2*ZH*ZH))
5124 GOTO 5170
5126 REM Geometrical lengths of sections
5128 REM Subroutine 5128
5130 LET W1=2* PI *F1
5132 FOR I=1 TO  INT (NN/2+.99)
5134 LET XQ=X(2*I-1)/ZH
5136 LET L(2*I-1)=VH/W1* ATN(XQ/SQR(1-XQ*XQ))
5138 NEXT I
5140 LET IT=1
5142 FOR I=1 TO  INT (NN/2)
5144 LET L(2*I)=(X(2*I)-2*CF*W1-W1/(2*ZH*VH)*(L(2*I-1)+L(2*I+1)
))*ZL*VL/W1
5146 NEXT I
5148 LET XQ=X(1)/ZH-ZL*W1*L(2)/(2*VL*ZH)
5150 LET L(1)= VH/W1*ATN(XQ/SQR(1-XQ*XQ))
5152 LET L(NN)=L(1)
5154 FOR I=2 TO  INT (NN/2)
5156 LET XQ=X(2*I-1)
5158 LET XQ=X(2*I-1)/ZH-ZL*W1/(2*ZH*VL)*(L(2*I-2)+L(2*I))
```

```
5160 LET L(2*I-1)=VH/W1*ATN(XQ/SQR(1-XQ*XQ))
5162 NEXT I
5164 LET IT=IT+1
5166 IF IT<3 THEN  GOTO 5142
5168 RETURN
5170 REM Chain subroutine
5172 PRINT
5174 INPUT"Enter 'CONT':",N$
5176 IF N$="CONT" THEN CLS:CHAIN "MENU",10
5178 GOTO 5172
5180 REM Control subroutine
5182 PRINT
5184 INPUT "Enter 'CONT':",N$
5186 IF N$="CONT" THEN CLS:RETURN
5188 GOTO 5182
```

5.7 MICROWAVE BANDPASS FILTERS

This section covers two common distributed element bandpass filters the electrical diagrams and microstrip layouts of which are shown in Figures 5.23 and 5.24. Filters of this type at microwave frequencies are widely used because they are easy to design and fabricate in the printed-circuit technique. This is especially true for filters with parallel-coupled resonators, such as those shown in Figure 5.23, when their fractional bandwidth is 15% or less. Under this constraint, these filters have reasonable, practical impedance levels. However, for larger bandwidths, the stub filter shown in Figure 5.24 is recommended [5].

In practice, the bandpass filters with Chebyschev frequency response are used most often. Therefore, the BPF computer program presented herein is intended for designing Chebyschev filters the frequency responses of which are similar to those presented in Figure 5.25.

We start the design process by introducing the following data into the computer memory:

Z_0 = the terminating resistances in ohms,
f_0 = the center frequency of the required passband in Hz,
w = the fractional bandwidth,
L_r = the maximum tolerable attenuation over the passband in dB,
f_a = the reference frequency in Hz,
L_a = the filter attenuation at frequency f_a, in dB,
I = when designing the filter shown in Figure 5.23, or
II = when designing the filter shown in Figure 5.24.

At the beginning, these input data are tested. If they are acceptable for design, the minimum number, n, of the reactive elements of the Chebyschev low-pass prototype filter is evaluated from

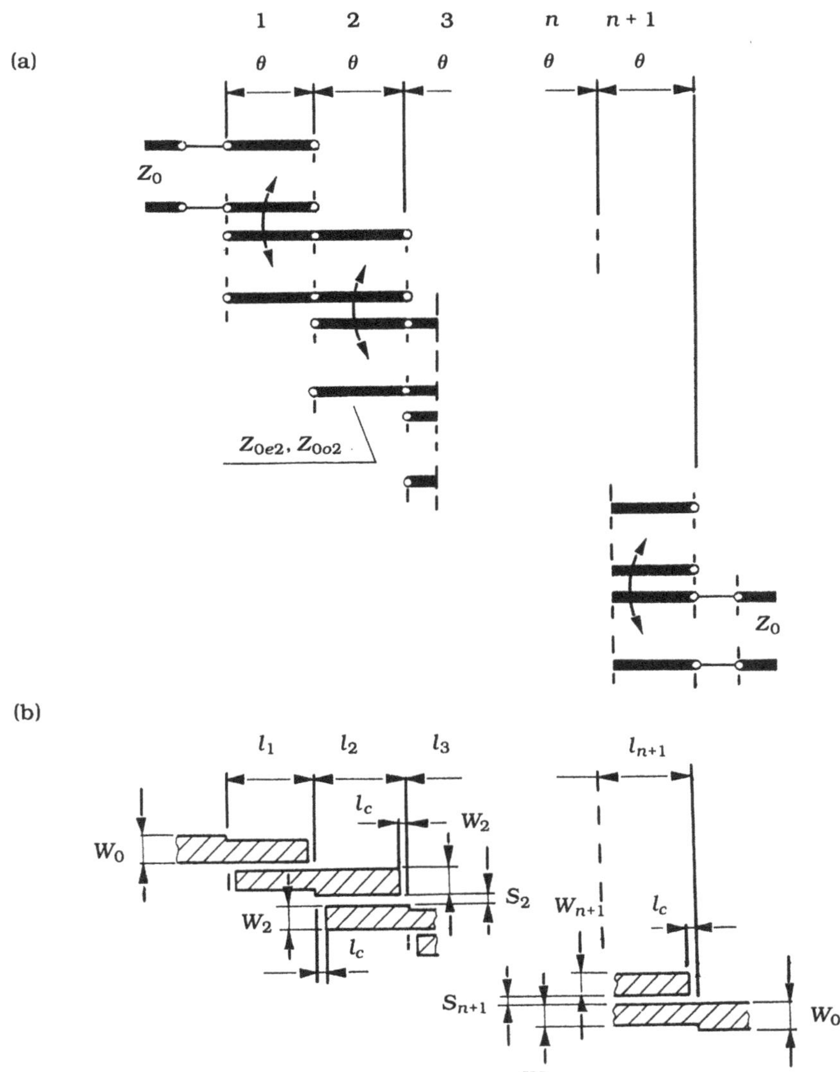

Figure 5.23 The microwave bandpass filter: (a) electrical scheme and (b) coupled microstrip lines configuration.

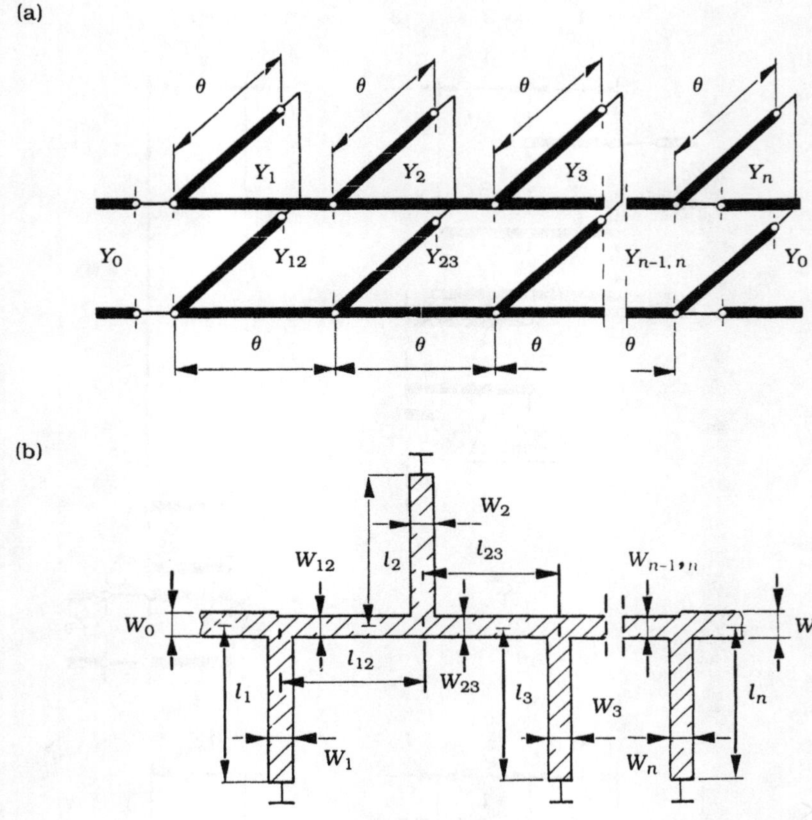

Figure 5.24 The microwave bandpass filter: (a) electrical scheme and (b) microstrip line configuration.

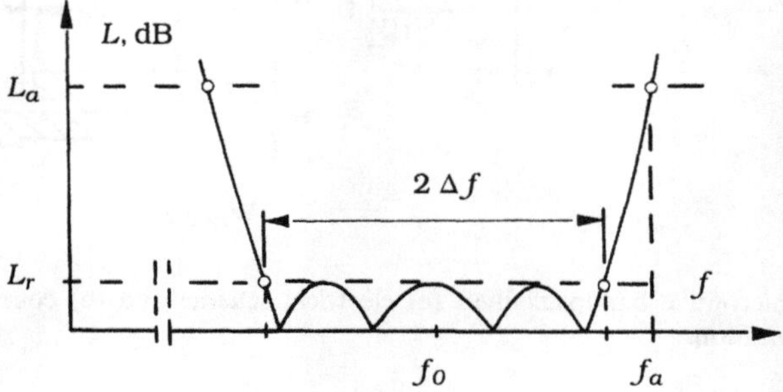

Figure 5.25 The typical insertion loss function for the Chebyschev bandpass filters.

$$n \geqslant \frac{\cosh^{-1}\sqrt{(L'_a - 1)/(L'_r - 1)}}{\cosh^{-1}(c_1/c_2)} \tag{5.51}$$

where $L'_a = 10^{(L_a/10)}$, $L'_r = 10^{(L_r/10)}$, $c_1 = \sin(\pi w_a/4)$, $c_2 = \sin(\pi w/4)$ and $w_a = 2(f_a - f_0)/f_0$.

Next, the elements g_k, where $k = 0, 1, 2, \ldots, n + 1$, of the Chebyschev low-pass prototype filter are calculated by using (5.46).

Cohn [18] has shown that even- and odd-mode impedances of particular sections of the filter presented in Figure 5.23 are related to elements g_k as follows:

$$\begin{aligned} Z_{0ei,i+1} &= \frac{1}{Y_0}\left[1 + \frac{J_{i,i+1}}{Y_0} + \left(\frac{J_{i,i+1}}{Y_0}\right)^2\right] \\ Z_{0oi,i+1} &= \frac{1}{Y_0}\left[1 - \frac{J_{i,i+1}}{Y_0} + \left(\frac{J_{i,i+1}}{Y_0}\right)^2\right] \end{aligned} \tag{5.52}$$

where

$$\begin{aligned} i &= 0, 1, 2, \ldots, n \\ Y_0 &= 1/Z_0 \\ J_{0,1} &= Y_0 \sqrt{\frac{\pi w}{2g_0g_1}} \\ J_{i,i+1} &= \frac{Y_0\pi w}{2\omega'_1}\frac{1}{\sqrt{g_ig_{i+1}}} \quad \text{for } i = 1, 2, \ldots, n-1 \\ J_{n,n+1} &= Y_0 \sqrt{\frac{\pi w}{2g_ng_{n+1}}} \\ \omega'_1 &= 1 \end{aligned}$$

The electrical lengths, θ, of all these sections are equal to each other, and become $\pi/2$ radians at the center band frequency f_0.

The filters composed of shunt stubs and connecting sections (see Figure 5.24) are designed in a similar way [5]. In other words, their electrical parameters are calculated on the basis of the suitable Chebyschev low-pass prototype filter. Therefore, in this case (5.51) and (5.46) are helpful. Next, the characteristic admittances Y_i of the shunt stubs and $Y_{i,i+1}$ of the connecting sections are calculated from

$$Y_1 = Y_0 \left[g_0(1 - d)g_1\omega_1' \tan(\theta_1) + N_{1,2} - \frac{I_{1,2}}{Y_0} \right]$$

$$Y_i = Y_0 \left(N_{i-1,i} + N_{i,i+1} - \frac{I_{i-1,i}}{Y_0} - \frac{I_{i,i+1}}{Y_0} \right) \quad (5.53)$$

for $i = 2, 3, \ldots, n - 1$ and

$$Y_n = Y_0\omega_1'(g_n g_{n+1} - dg_0 g_1)\tan(\theta_1) + Y_0 \left(N_{n-1,n} - \frac{I_{n-1,n}}{Y_0} \right)$$

$$Y_{i,i+1} = Y_0 \left(\frac{I_{i,i+1}}{Y_0} \right) \quad \text{for } i = 1, 2, \ldots, n - 1$$

$$\frac{I_{1,2}}{Y_0} = g_0 \sqrt{\frac{C_a}{g_2}}$$

$$\frac{I_{k,k+1}}{Y_0} = \frac{g_0 C_a}{\sqrt{g_k g_{k+1}}} \quad \text{for } k = 2, 3, \ldots, n - 2$$

$$\frac{I_{n-1,n}}{Y_0} = g_0 \sqrt{\frac{C_a g_{n+1}}{g_0 g_{n-1}}}$$

$$N_{k,k+1} = \sqrt{\left(\frac{I_{k,k+1}}{Y_0} \right)^2 + \left[\frac{g_0\omega_1' C_a \tan(\theta_1)}{2} \right]^2}$$

where $\omega_1' = 1$, $\theta_1 = \pi(2 - w)/4$, $C_a = 2g_1 d$ and d is the dimensional constant, which can be chosen to give a convenient admittance level in the interior of the filter, usually d is chosen as 1.

Here we should remember that all shunt stubs, short-circuited at their ends, and connecting sections (see Figure 5.24) are $\lambda_0/4$ long, where λ_0 is the wavelength in the medium of propagation at the center band frequency f_0. By knowing the electrical parameters, we can evaluate some geometrical dimensions of the filter being designed. This can be done automatically when the filter is constructed of segments of single and coupled transmission lines, as indicated in Figure 5.26. For this purpose, we must run the program BPF again by using the command CONTINUE, and then introduce the suitable structural data requested by the computer. Short descriptions of these structural data are given in Chapter 6. Finally, we obtain the main geometrical filter dimensions, as shown in Figures 5.23(b) and 5.24(b).

Check Calculations

The use of the BPF computer program is illustrated by the following examples.

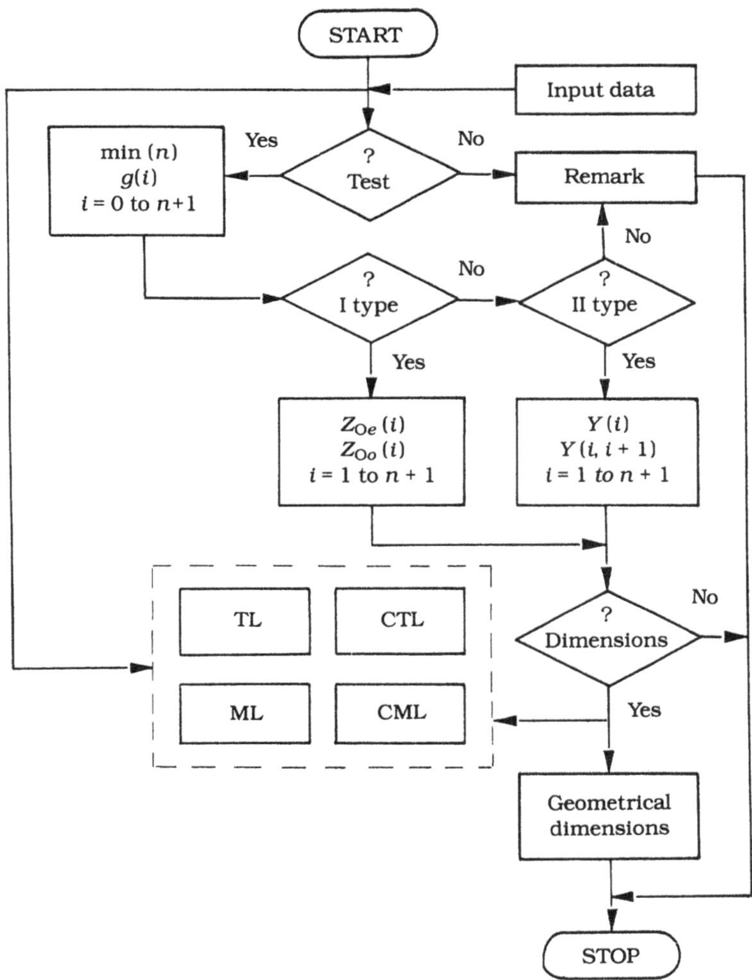

Figure 5.26 The flow diagram of the BPF computer program.

Example 1

Design the Chebyschev bandpass filter, as shown in Figure 5.23, with the following parameters: $Z_0 = 50\ \Omega, f_0 = 3\ 10^9$ Hz, $w = 0.1$, $L_r = 0.1$ dB, $f_a = 3.3\ 10^9$ Hz and $L_a = 30$ dB. Assume that the filter will be constructed of segments of the single and coupled striplines (see Sections 6.4 and 6.8), which are characterized by $\epsilon_r = 2.65$, $\mu_r = 1$, $b = 3.52\ 10^{-3}$ m and $t \ll b$.

Table 5.7 presents some electrical and structural filter parameters obtained for the above given data. The dimension l_c shown in Figure 5.23(b) is a resonator length correction to account for the fringing capacitance from the end of each strip. Cohn [18] has found that a constant correction of $l_c = 0.165\ b$, where b is the dielectric substrate thickness, is apparently satisfactory for many applications.

Table 5.7
Electrical Parameters and Geometrical Dimensions of the Microwave Bandpass Filter Shown in Figure 5.23

	$n = 5,\ W_0 = 2.52\ 10^{-3}\ m$				
i	Z_{0ei}, ohm	Z_{0oi}, ohm	W_i, m	S_i, m	l_i, m
1	75.35	38.34	$1.82\ 10^{-3}$	$0.21\ 10^{-3}$	$15.35\ 10^{-3}$
2	57.04	44.52	$2.41\ 10^{-3}$	$0.96\ 10^{-3}$	$15.35\ 10^{-3}$
3	55.22	45.68	$2.45\ 10^{-3}$	$1.23\ 10^{-3}$	$15.35\ 10^{-3}$
4	55.22	45.68	$2.45\ 10^{-3}$	$1.23\ 10^{-3}$	$15.35\ 10^{-3}$
5	57.04	44.52	$2.41\ 10^{-3}$	$0.96\ 10^{-3}$	$15.35\ 10^{-3}$
6	75.35	38.34	$1.82\ 10^{-3}$	$0.21\ 10^{-3}$	$15.35\ 10^{-3}$

Example 2

Design the Chebyschev bandpass filter shown in Figure 5.24, satisfying the following requirements: $Z_0 = 50\ \Omega$, $f_0 = 10^9$ Hz, $w = 0.7$, $L_r = 0.1$ dB, $f_a = 1.6\ 10^9$ Hz and $L_a = 45$ dB. Assume that the filter being designed is composed of segments of microstrip line (see Section 6.5) with $\epsilon_r = 4.23$, $\mu_r = 1$, $h = 10^{-3}$ m and $t = 3.5\ 10^{-5}$ m. Some electrical and structural filter parameters obtained for the assumed above input data are summarized in Table 5.8.

Table 5.8
Electrical Parameters and Geometrical Dimensions of the Microwave Bandpass Filter Shown in Figure 5.24

	$n = 8,\ W_0 = 1.93\ 10^{-3}\ m$					
i	Z_i, ohm	$Z_{i,i+1}$, ohm	W_i, m	$W_{i,i+1}$, m	l_i, m	$l_{i,i+1}$, m
1	47.986	38.823	$2.06\ 10^{-3}$	$2.87\ 10^{-3}$	$41.60\ 10^{-3}$	$41.00\ 10^{-3}$
2	24.383	36.644	$5.48\ 10^{-3}$	$3.13\ 10^{-3}$	$39.80\ 10^{-3}$	$40.84\ 10^{-3}$
3	24.403	38.710	$5.47\ 10^{-3}$	$2.89\ 10^{-3}$	$39.80\ 10^{-3}$	$40.99\ 10^{-3}$
4	23.954	39.165	$5.61\ 10^{-3}$	$2.84\ 10^{-3}$	$39.76\ 10^{-3}$	$41.02\ 10^{-3}$
5	23.954	38.710	$5.61\ 10^{-3}$	$2.89\ 10^{-3}$	$39.76\ 10^{-3}$	$40.99\ 10^{-3}$
6	24.403	36.644	$5.47\ 10^{-3}$	$3.13\ 10^{-3}$	$39.80\ 10^{-3}$	$40.84\ 10^{-3}$
7	24.383	38.823	$5.48\ 10^{-3}$	$2.87\ 10^{-3}$	$39.80\ 10^{-3}$	$41.00\ 10^{-3}$
8	47.986	—	$2.06\ 10^{-3}$	—	$41.60\ 10^{-3}$	—

Listing of Program BPF

```
5300 REM BPF
5302 CLS
5304 CHAIN MERGE "ML",5306
5306 CHAIN MERGE "TL",5308
5308 CHAIN MERGE "CTL",5310
5310 PRINT "BAND-PASS FILTERS"
5312 PRINT
5314 PRINT "Data:"
5316 PRINT
5318 INPUT "Zo[ohm]=",ZO
5320 INPUT "fo[Hz]=",FO
5322 INPUT "w=",W
5324 INPUT "Lr[dB]=",LR
5326 INPUT "fa[Hz]=",FA
5328 INPUT "La[dB]=",LA
5330 LET PI=3.141593
5332 PRINT
5334 INPUT "Enter the name  ( I - Fig. 5.23 or II - Fig. 5.24 )
 of the filter to be designed:",S$
5336 IF S$="I" OR S$="II" THEN  GOTO 5342
5338 PRINT "Wrong command !"
5340 GOTO 5332
5342 PRINT
5344 IF FA>FO*(1+W/2) THEN  GOTO 5348
5346 PRINT "Error:fa <= fo*(1+w/2)": GOTO 5312
5348 IF ZO <= 0 THEN  PRINT "Error:Zo <= 0": GOTO 5316
5350 IF W <= 0 THEN  PRINT "Error:w <= 0": GOTO 5316
5352 IF W >= 2 THEN  PRINT "Error:w >= 2": GOTO 5316
5354 IF W >= 2 THEN  PRINT "Error:w >= 2": GOTO 5316
5356 IF LR <= 0 THEN  PRINT "Error:Lr <= 0": GOTO 5316
5358 IF LA <= LR THEN  PRINT "Error:La <= Lr": GOTO 5316
5360 LET X= SQR ((10^(LA/10)-1)/(10^(LR/10)-1))
5362 LET A= LOG(X+ SQR (X*X-1))
5364 LET X= SIN ( PI *(FA-FO)/(FO*2))/ SIN ( PI *W/4)
5366 LET B= LOG(X+ SQR (X*X-1))
5368 LET N= INT (.99+A/B)
5370 DIM A(N)
5372 DIM B(N)
5374 DIM G(N+1)
5376 REM Calculation of g(i)
5378 LET X= EXP (LR/17.37)
5380 LET C= LOG((X+1/X)/(X-1/X))
5382 LET X= EXP (C/(2*N))
5384 LET Y=(X-1/X)/2
5386 FOR I=1 TO N
5388 LET A(I)= SIN ((2*I-1)* PI /(2*N))
5390 LET B(I)=Y*Y+ SIN (I* PI /N)* SIN (I* PI /N)
5392 NEXT I
5394 LET G(1)=2*A(1)/Y
5396 FOR I=2 TO N
5398 LET G(I)=4*A(I-1)*A(I)/(B(I-1)*G(I-1))
5400 NEXT I
```

```
5402 IF (N/2- INT (N/2))>.05 THEN  GOTO 5410
5404 LET X= EXP (C/4)
5406 LET G(N+1)=((X+1/X)/(X-1/X))^2
5408 GOTO 5412
5410 LET G(N+1)=1
5412 IF S$="I" THEN  GOTO 5418
5414 IF S$="II" THEN  GOTO 5446
5416 REM Calculation of Zoe(i) and  Zoo(i)
5418 DIM J(N+1)
5420 LET J(1)= SQR ( PI *W/(2*G(1)))
5422 FOR I=2 TO N
5424 LET J(I)= PI *W/(2* SQR (G(I-1)*G(I)))
5426 NEXT I
5428 LET J(N+1)= SQR ( PI *W/(2*G(N)*G(N+1)))
5430 DIM E(N+1)
5432 DIM O(N+1)
5434 FOR I=1 TO N+1
5436 LET E(I)=ZO*(1+J(I)+J(I)*J(I))
5438 LET O(I)=ZO*(1-J(I)+J(I)*J(I))
5440 NEXT I
5442 GOTO 5490
5444 REM Calculation of Y(i) and Y(i,i+1)
5446 DIM J(N-1)
5448 LET O1= PI *(1-W/2)/2
5450 LET C=2*G(1)
5452 LET J(1)= SQR (C/G(2))
5454 FOR I=2 TO N-2
5456 LET J(I)=C/ SQR (G(I)*G(I+1))
5458 NEXT I
5460 LET J(N-1)= SQR (C*G(N+1)/G(N-1))
5462 FOR I=1 TO N-1
5464 LET A(I)= SQR (J(I)*J(I)+(C* TAN (O1)/2)^2)
5466 NEXT I
5468 DIM R(N)
5470 DIM S(N)
5472 LET R(1)=(A(1)-J(1))/ZO
5474 FOR I=2 TO N-1
5476 LET R(I)=(A(I-1)+A(I)-J(I-1)-J(I))/ZO
5478 NEXT I
5480 LET R(N)=(G(N)*G(N+1)-G(1))/ZO* TAN (O1)+(A(N-1)-J(N-1))/Z
O
5482 FOR I=1 TO N-1
5484 LET S(I)=J(I)/ZO
5486 NEXT I
5488 PRINT
5490 PRINT "Electrical parameters"
5492 PRINT
5494 PRINT "n=";N
5496 PRINT
5498 PRINT "Impedances in ohms"
5500 PRINT
5502 IF S$="I" THEN  GOTO 5522
5504 FOR I=1 TO N
5506 PRINT "Z(";I;")=";1/R(I)
5508 NEXT I
```

```
5510 PRINT
5512 FOR I=1 TO N-1
5514 PRINT "Z(";I;I+1;")=";1/S(I)
5516 NEXT I
5518 GOSUB 5720
5520 GOTO 5532
5522 FOR I=1 TO N+1
5524 PRINT "Zoe(";I;")=";E(I)
5526 PRINT "Zoo(";I;")=";O(I)
5528 NEXT I
5530 GOSUB 5720
5532 PRINT
5534 REM Geometrical dimensions
5536 IF S$="I" THEN  GOTO 5660
5538 INPUT "Enter the name ( TL or ML ) of the realization tech
nique:",H$
5540 PRINT
5542 IF H$="TL" THEN  GOTO 5546
5544 IF H$="ML" THEN  GOTO 5602
5546 PRINT
5548 PRINT "Stripline realization"
5550 PRINT
5552 INPUT "er=",ER
5554 INPUT "ur=",UR
5556 INPUT "b[m]=",B
5558 INPUT "t[m]=",T
5560 PRINT
5562 PRINT "Geometrical dimensions in meters, se Fig. 5.24(b)"
5564 PRINT
5566 FOR J=1 TO N
5568 LET ZC=1/R(J)
5570 GOSUB 8252
5572 PRINT "W(";J;")=";W
5574 PRINT "l(";J;")=";75*10^6/(FO* SQR (ER))
5576 NEXT J
5578 GOSUB 5720
5580 FOR J=1 TO N-1
5582 LET ZC=1/S(J)
5584 GOSUB 8252
5586 PRINT "W(";J;J+1;")=";W
5588 PRINT "l(";J;J+1;"=";75*10^6/(FO* SQR (ER))
5590 NEXT J
5592 LET ZC=ZO
5594 GOSUB 8252
5596 PRINT
5598 PRINT "Wo=";W
5600 GOTO 5710
5602 PRINT
5604 PRINT "Microstrip line realization"
5606 PRINT
5608 INPUT "er=",ER
5610 INPUT "h[m]=",H
5612 INPUT "t[m]=",T
5614 LET F=FO
5616 LET NN=N
```

```
5618 PRINT
5620 PRINT "Geometrical dimensions in meters, see Fig. 5.24(b)"
5622 PRINT
5624 FOR J=1 TO NN
5626 LET ZC=1/R(J)
5628 GOSUB 8454
5630 PRINT "W(";J;")=";(U-DTU)*H
5632 PRINT "l(";J;")=";75*10^6/(FO* SQR (EF))
5634 NEXT J
5636 GOSUB 5720
5638 FOR J=1 TO NN-1
5640 LET ZC=1/S(J)
5642 GOSUB 8454
5644 PRINT "W(";J;J+1;")=";(U-DTU)*H
5646 PRINT "l(";J;J+1;")=";75*10^6/(FO* SQR ( EF))
5648 NEXT J
5650 LET ZC=ZO
5652 GOSUB 8454
5654 PRINT
5656 PRINT "Wo=";(U-DTU)*H
5658 GOTO 5710
5660 PRINT
5662 PRINT "Coupled striplines realization"
5664 PRINT
5666 INPUT "er=",ER
5668 INPUT "ur=",UR
5670 INPUT "b[m]=",B
5672 INPUT "t[m]=",T
5674 PRINT
5676 PRINT "Geometrical dimensions in meters, see Fig. 5.23(b).
"
5678 PRINT
5680 FOR J=1 TO N+1
5682 LET ZCE=E(J)
5684 LET ZCO=O(J)
5686 GOSUB 9360
5688 PRINT "W(";J;")=";W
5690 PRINT "S(";J;")=";S
5692 PRINT "l(";J;")=";75*10^6/(FO* SQR (ER))
5694 PRINT
5696 IF (J MOD 5)=0 THEN PRINT"Press any key" ELSE GOTO 5700
5698 IF INKEY$="" THEN GOTO 5698 ELSE CLS
5700 NEXT J
5702 LET ZC=ZO
5704 GOSUB 8252
5706 PRINT
5708 PRINT "Wo=";W
5710 REM Chain subroutine
5712 PRINT
5714 INPUT"Enter 'CONT':",N$
5716 IF N$="CONT" THEN CLS:CHAIN "MENU",10
5718 GOTO 5712
5720 REM Control subroutine
5722 PRINT
```

```
5724 INPUT "Enter 'CONT':",N$
5726 IF N$="CONT" THEN CLS:RETURN
5728 GOTO 5722
```

5.8 MICROWAVE BANDSTOP FILTERS

This section presents explicit formulas for the design of distributed bandstop filters the general electrical scheme and microstrip outline of which are shown in Figure 5.27. The insertion loss function of these filters is presented in Figure 5.28. As we

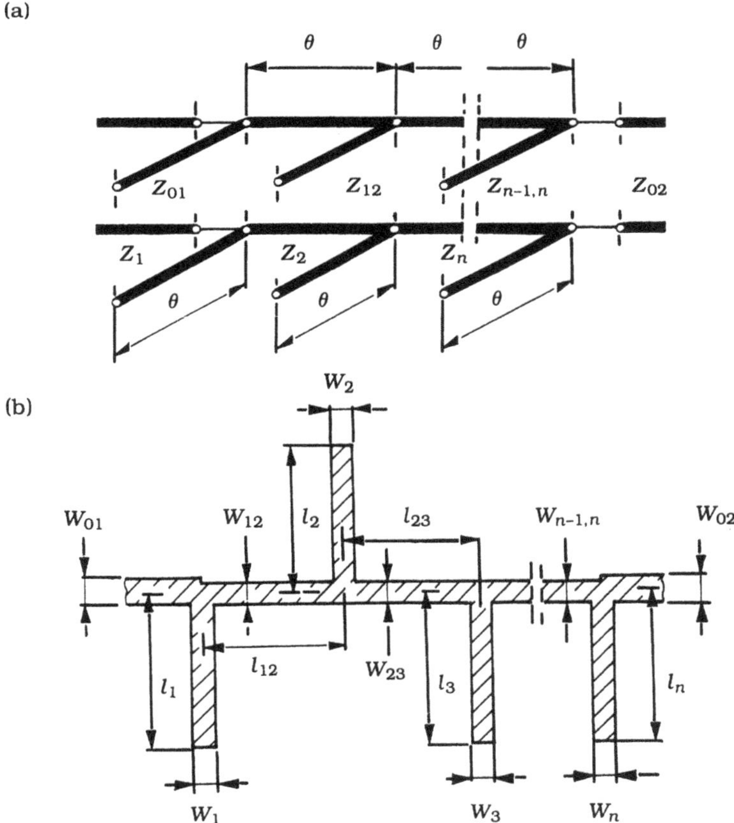

Figure 5.27 The microwave bandstop filter: (a) electrical scheme and (b) microstrip line configuration.

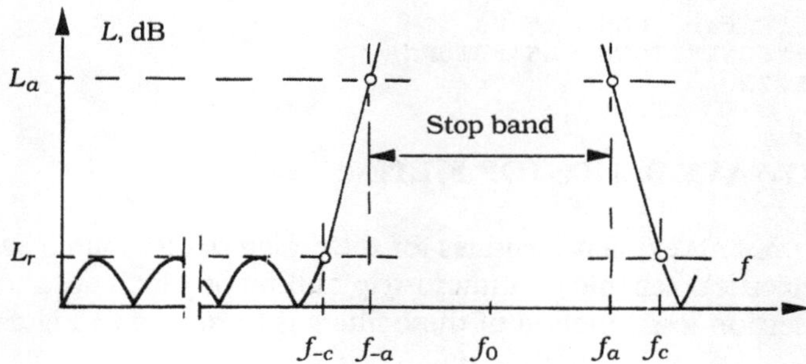

Figure 5.28 The typical insertion loss function for the Chebyschev bandstop filters.

can see, frequencies f_{-c} and f_c determine the lower and upper limits of the first stop band at attenuation level L_r. Similarly, frequencies f_{-a} and f_a determine the width of this band at level L_a.

In general, the filter of degree n is composed of $2n - 1$ commensurate transmission line segments the electrical lengths θ of which are equal to each other and become $\pi/2$ radians at the center band frequency f_0. Here, we note that the filter presented is of particular interest because in most cases it can be realized without difficulty using the printed-circuit technique. Moreover, its design is not complicated due to the closed-form relations derived in [19]. According to [19], as well as [5], the design process starts with the evaluation of the minimum value n of the filter degree from the following criterion:

$$n \geqslant \frac{\cosh^{-1}\sqrt{(L'_a - 1)/(L'_r - 1)}}{\cosh^{-1}(c_1/c_2)} \tag{5.54}$$

where $L'_a = 10^{(L_a/10)}$, $L'_r = 10^{(L_r/10)}$, $f_0 = (f_{-c} + f_c)/2$, $c_1 = \tan[\pi f_{-a}/(2f_0)]$ and $c_2 = \tan[\pi f_{-c}/(2f_0)]$.

Next, the characteristic impedances Z_i of open-circuited shunt stubs and impedances $Z_{j-1,j}$ of connecting sections (see Figure 5.27) are calculated by using the formulas presented below.

(1) $n = 1$

$$Z_1 = Z_{01}/(dg_0g_1) \tag{5.55}$$

$$Z_{02} = Z_{01}g_2/g_0$$

(2) $n = 2$

$$
\begin{aligned}
Z_1 &= Z_{01}[1 + 1/(dg_0g_1)] \\
Z_{12} &= Z_{01}(1 + dg_0g_1) \qquad (5.56) \\
Z_2 &= Z_{01}g_0/(dg_2) \\
Z_{02} &= Z_{01}g_0g_3
\end{aligned}
$$

(3) $n = 3$

The impedances Z_1, Z_2 and Z_{12} are calculated by using the formulas given in point 2 (i.e., for $n = 2$).

$$
\begin{aligned}
Z_3 &= Z_{01}g_0[1 + 1/(dg_3g_4)]/g_4 \\
Z_{23} &= Z_{01}g_0(1 + dg_3g_4)/g_4 \qquad (5.57) \\
Z_{02} &= Z_{01}g_0/g_4
\end{aligned}
$$

(4) $n = 4$

$$
\begin{aligned}
Z_1 &= Z_{01}[2 + 1/(dg_0g_1)] \\
Z_{12} &= Z_{01}(1 + 2dg_0g_1)/(1 + dg_0g_1) \\
Z_2 &= Z_{01}/(1 + dg_0g_1) + Z_{01}g_0/[dg_2(1 + dg_0g_1)^2] \\
Z_{23} &= Z_{01}[dg_2 + g_0/(1 + dg_0g_1)]/g_0 \qquad (5.58) \\
Z_3 &= Z_{01}/(dg_0g_1) \\
Z_{34} &= Z_{01}(1 + dg_4g_5)/(g_0g_5) \\
Z_4 &= Z_{01}[1 + 1/(dg_4g_5)]/(g_0g_5) \\
Z_{02} &= Z_{01}/(g_0g_5)
\end{aligned}
$$

(5) $n = 5$

The impedances Z_1, Z_2, Z_3, Z_{12} and Z_{23} are calculated by using the corresponding formulas given in point 4 (i.e., for $n = 4$).

$$
\begin{aligned}
Z_4 &= Z_{01}/[g_0(1 + dg_5g_6)] + Z_{01}g_6/[dg_0g_4(1 + g_4g_5)^2] \\
Z_{34} &= Z_{01}[dg_4 + g_6/(1 + dg_5g_6)]/g_0 \\
Z_5 &= Z_{01}g_6[2 + 1/(dg_5g_6)]/g_0 \qquad (5.59) \\
Z_{45} &= Z_{01}g_6(1 + 2dg_5g_6)/[g_0(1 + dg_5g_6)] \\
Z_{02} &= Z_{01}g_6/g_0
\end{aligned}
$$

In all the above presented design formulas, the parameter $d = \cot[\pi f_{-c}/(2f_0)]$ and $g_0, g_1, g_2, \ldots, g_{n+1}$ are elements of the Chebyschev low-pass prototype filter defined by (5.46).

Formulas (5.54) to (5.59) have been used in the computer program BSF, the flow diagram of which is shown in Figure 5.29. The input data for this program are the values of Z_{01}, f_{-c}, f_c, f_{-a}, L_r and L_a the meaning of which is given in Figure 5.28 and at the beginning of this section.

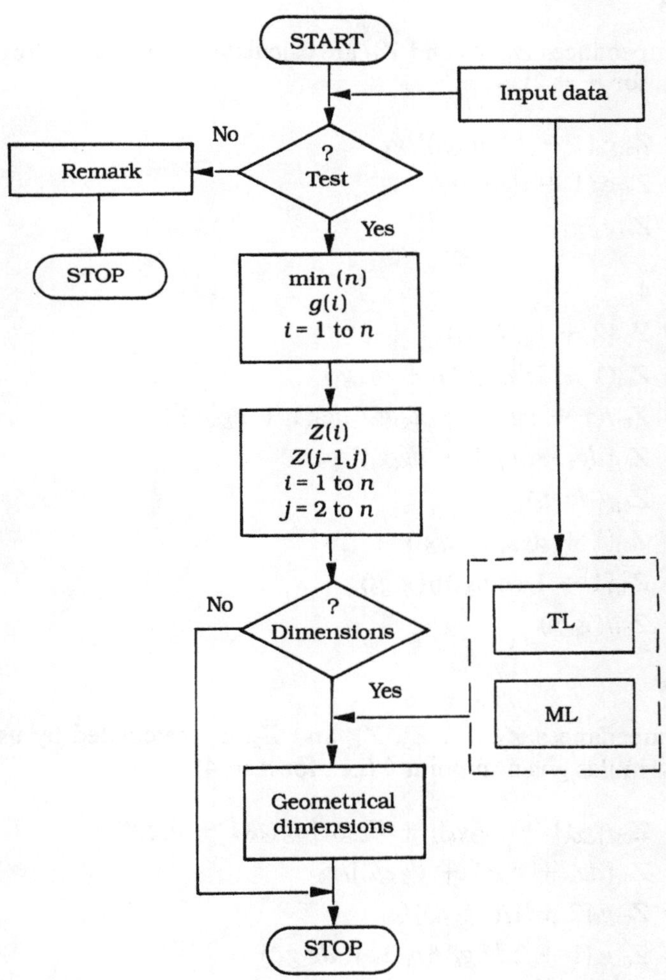

Figure 5.29 The flow diagram of the BSF computer program.

After calculating the electrical parameters of the filter being designed, we may evaluate some of its geometrical dimensions. This can be done automatically when the filter is constructed of segments of the triplate (TL) or microstrip (ML) lines. For this purpose, we run the program BSF again (by using the command CONTINUE), and then introduce the suitable structural data requested by the computer. Short descriptions of these structural data are given in Sections 6.4 and 6.5. As output results, we obtain main geometrical dimensions of the filter, as shown in Figure 5.27(b). In the design process, only lengths of the open-circuited shunt stubs are corrected (shortened) for the fringing capacitances from their ends. The length corrections l_{ci}, where $i = 1, 2, \ldots, n$, are calculated by using the following approximate formulas [11,17].

For the triplate stripline shunt stubs:

$$l_{ci} = \frac{\lambda}{2\pi} \arctan\left[\frac{q + 2W_i}{4q + 2W_i} \tan(2\pi q/\lambda)\right] \tag{5.60}$$

where $q = b \ln(2)/\pi$, b is the dielectric substrate thickness, W_i is the width of the ith shunt stub, and λ is a wavelength in the medium of propagation. For the microstrip line shunt stubs:

$$l_{ci} = 0.412h\left(\frac{\epsilon_{efi} + 0.3}{\epsilon_{efi} - 0.258}\right)\left(\frac{W_i/h + 0.263}{W_i/h + 0.813}\right) \tag{5.61}$$

where h is the dielectric substrate thickness, ϵ_{efi} is the effective dielectric constant of the ith shunt stub, and W_i is the width of this stub.

Check Calculations

Computer program BSF has been tested for the following data:

$Z_{01} = 30$ Ω

$f_{-c} = 1.12 \cdot 10^9$ Hz

$f_c = 2.08 \cdot 10^9$ Hz

$f_{-a} = 1.48 \cdot 10^9$ Hz

$L_r = 0.1$ dB

$L_a = 30$ dB

We have assumed that the filter is constructed of segments of the microstrip line (see Figure 6.4) with

$$\epsilon_r = 4.23$$
$$h = 1 \cdot 10^{-3} \text{ m}$$
$$t = 35 \cdot 10^{-6} \text{ m}$$

The computed values of the filter impedances are given in the second and third columns of Table 5.9. This table also presents some geometrical filter dimensions. In the design process, geometrical lengths of the shunt stubs (see Figure 5.27) have been corrected in accordance with (5.61).

Table 5.9
Electrical Parameters and Geometrical Dimensions of the Microwave Bandstop Filter Shown in Figure 5.27

$n = 3$, $Z_{01} = Z_{02} = 30\ \Omega$, $W_{01} = W_{02} = 4.15\ 10^{-3}$ m

i	Z_i, ohm	$Z_{i,i+1}$, ohm	W_i, m	$W_{i,i+1}$, m	$l_i - l_{ci}$, m	$l_{i,i+1}$, m
1	87.075	45.768	$0.616\ 10^{-3}$	$2.225\ 10^{-3}$	$26.78\ 10^{-3}$	$25.89\ 10^{-3}$
2	51.314	45.768	$1.841\ 10^{-3}$	$2.225\ 10^{-3}$	$25.71\ 10^{-3}$	$25.89\ 10^{-3}$
3	87.075	—	$0.616\ 10^{-3}$	—	$26.78\ 10^{-3}$	—

Listing of Program BSF

```
6200 REM BSF
6202 CLS
6204 PRINT "BAND-STOP FILTERS"
6206 CHAIN MERGE "ML",6208
6208 PRINT
6210 PRINT "Data:"
6212 PRINT
6214 INPUT "Zo1[ohm]=",ZO1
6216 INPUT "f(-c)[Hz]=",F1
6218 INPUT "f(c)[Hz]=",F2
6220 INPUT "f(-a)[Hz]=",F3
6222 INPUT "Lr[dB]=",LR
6224 INPUT "La[dB]=",LA
6226 LET PI=3.141593
6228 PRINT
6230 IF ZO1 <= 0 THEN  PRINT "Error:Zo1 <= 0": GOTO 6212
6232 IF F3 <= F1 THEN  PRINT "Error:f(-c) >= f(-a)": GOTO 6212
6234 IF LR <= 0 THEN  PRINT "Error:Lr <= 0": GOTO 6212
6236 IF LA <= LR THEN  PRINT "Error:La <= Lr": GOTO 6212
6238 LET C1=(10^(LA/10)-1)/(10^(LR/10)-1)
6240 LET C2=1/ TAN ( PI *F1/(F1+F2))* TAN ( PI *F3/(F1+F2))
```

```
6242 LET A= SQR (C1)
6244 LET B=C2
6246 LET L= LOG(A+ SQR (A*A-1))
6248 LET M= LOG(B+ SQR (B*B-1))
6250 LET N= INT (.99+L/M)
6252 IF N <= 5 THEN  GOTO 6258
6254 PRINT "n>5 !"
6256 GOTO 6448
6258 DIM A(N)
6260 DIM B(N)
6262 DIM G(N+1)
6264 DIM R(N)
6266 DIM S(N)
6268 LET X= EXP (LR/17.37)
6270 LET C= LOG((X+1/X)/(X-1/X))
6272 LET X= EXP (C/(2*N))
6274 LET Y=(X-1/X)/2
6276 FOR I=1 TO N
6278 LET A(I)= SIN ((2*I-1)* PI /(2*N))
6280 LET B(I)=Y*Y+ SIN (I* PI /N)* SIN (I* PI /N)
6282 NEXT I
6284 LET G(1)=2*A(1)/Y
6286 FOR I=2 TO N
6288 LET G(I)=4*A(I-1)*A(I)/(B(I-1)*G(I-1))
6290 NEXT I
6292 IF (N/2- INT (N/2))>.05 THEN  LET G(N+1)=1: GOTO 6298
6294 LET X= EXP (C/4)
6296 LET G(N+1)=((X+1/X)/(X-1/X))^2
6298 LET D=1/ TAN ( PI *F1/(F1+F2))
6300 IF N=1 THEN  LET R(1)=ZO1/(D*G(1)): LET ZO2=ZO1*G(2): GOTO
 6350
6302 IF N>3 THEN  GOTO 6322
6304 LET R(1)=ZO1*(1+1/(D*G(1)))
6306 LET R(2)=ZO1/(D*G(2))
6308 LET S(1)=ZO1*(1+D*G(1))
6310 LET ZO2=ZO1*G(3)
6312 IF N=2 THEN  GOTO 6350
6314 LET R(3)=ZO1/G(4)*(1+1/(D*G(3)*G(4)))
6316 LET S(2)=ZO1/G(4)*(1+D*G(3)*G(4))
6318 LET ZO2=ZO1/G(4)
6320 GOTO 6350
6322 LET R(1)=ZO1*(2+1/(D*G(1)))
6324 LET S(1)=ZO1*((1+2*D*G(1))/(1+D*G(1)))
6326 LET R(2)=ZO1*(1/(1+D*G(1))+1/(D*G(2)*(1+D*G(1))^2))
6328 LET S(2)=ZO1*(D*G(2)+1/(1+D*G(1)))
6330 LET R(3)=ZO1/(D*G(1))
6332 LET S(3)=ZO1/G(5)*(1+D*G(4)*G(5))
6334 LET R(4)=ZO1/G(5)*(1+1/(D*G(4)*G(5)))
6336 LET ZO2=ZO1/G(5)
6338 IF N=4 THEN  GOTO 6350
6340 LET R(4)=ZO1*(1/(1+D*G(5)*G(6))+G(6)/(D*G(4)*(1+G(4)*G(5))
^2))
6342 LET S(3)=ZO1*(D*G(4)+G(6)/(1+D*G(5)*G(6)))
6344 LET R(5)=ZO1*G(6)*(2+1/(D*G(5)*G(6)))
6346 LET S(4)=ZO1*G(6)*(1+2*D*G(5)*G(6))/(1+D*G(5)*G(6))
```

```
6348 LET ZO2=ZO1*G(6)
6350 PRINT "Results, see Fig. 5.27"
6352 PRINT
6354 PRINT "n = ";N
6356 PRINT
6358 PRINT "Zo1[ohm]=";ZO1
6360 PRINT "Zo2[ohm]=";ZO2
6362 PRINT
6364 LET NN=N
6366 FOR J=1 TO NN
6368 PRINT "Z(";J;")[ohm]=";R(J)
6370 NEXT J
6372 PRINT
6374 FOR J=1 TO NN-1
6376 PRINT "Z(";J;J+1;")[ohm]=";S(J)
6378 NEXT J
6380 GOSUB 6458
6382 PRINT
6384 PRINT "Microstrip line realization:"
6386 PRINT
6388 INPUT "er=",ER
6390 INPUT "h[m]=",H
6392 INPUT "t[m]=",T
6394 PRINT
6396 LET FC=.5*(F1+F2)
6398 PRINT "Geometrical dimensions in meters, see Fig.5.27(b)"
6400 PRINT
6402 PRINT "Please wait !"
6404 PRINT
6406 LET ZC=ZO1
6408 GOSUB 8454
6410 PRINT "Wo1 =";(U-DTU)*H
6412 LET ZC=ZO2
6414 GOSUB 8454
6416 PRINT "Wo2 =";(U-DTU)*H
6418 PRINT
6420 FOR J=1 TO NN
6422 LET ZC=R(J)
6424 GOSUB 8454
6426 PRINT "W(";J;")=";(U-DTU)*H
6428 LET DL=.412*H*(EF+.3)*(U+.263)/((EF-.258)*(U+.813))
6430 PRINT "l(";J;")=";75*10^6/( SQR (EF)*FC)-DL
6432 NEXT J
6434 PRINT
6436 FOR J=1 TO NN-1
6438 LET ZC=S(J)
6440 GOSUB 8454
6442 PRINT "W(";J;J+1;")=";(U-DTU)*H
6444 PRINT "l(";J;J+1;")=";75*10^6/( SQR (EF)*FC)
6446 NEXT J
6448 REM Chain subroutine
6450 PRINT
6452 INPUT"Enter 'CONT':",N$
6454 IF N$="CONT" THEN CLS:CHAIN "MENU",10
```

```
6456 GOTO 6450
6458 REM Control subroutine
6460 PRINT
6462 INPUT "Enter 'CONT':",N$
6464 IF N$="CONT" THEN CLS:RETURN
6466 GOTO 6460
```

5.9 THE MULTIHARMONIC FILTER

Figure 5.30(a) shows the electrical scheme of a multiharmonic filter. According to the notation of this scheme, the filter is composed of the lumped capacitor C and the three-section transmission line short-circuited at its end.

In general, the driving point impedance of the stepped transmission line (see Figure 5.30(b)), can be expressed in terms of the signal frequency f, one-section

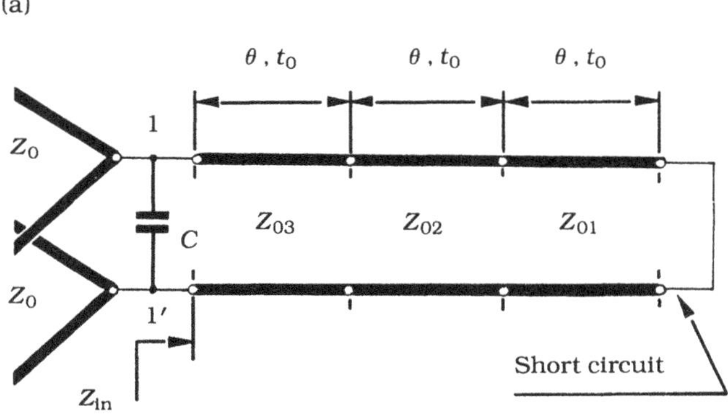

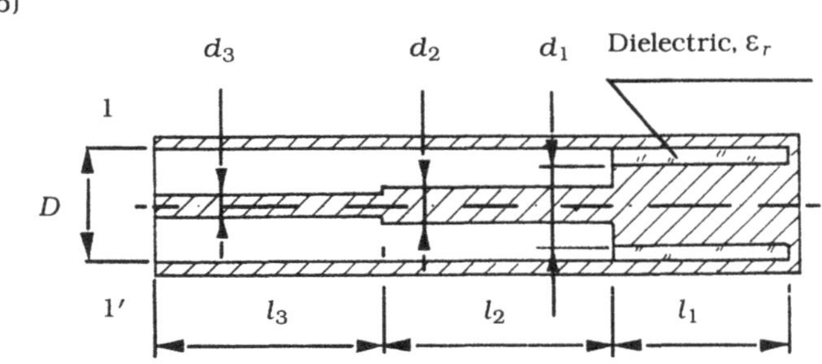

Figure 5.30 The multiharmonic bandstop filter: (a) electrical scheme and (b) coaxial line configuration of the stepped line short-circuited at its end.

delay time t_0 and characteristic impedances Z_{01}, Z_{02} and Z_{03} of the sections. If we choose suitable parameters, we make possible the design of a filter that ensures a current resonance at fundamental frequency f_1 and extremely low values of impedance $Z_{in}(f)$ at harmonic frequencies, mainly $f_2 = 2f_1$ and $f_3 = 3f_1$. To design the filter with such frequency response (see Figure 5.31), we must evaluate the appropriate values of t_0, Z_{01}, Z_{02}, Z_{03} and C. For this purpose, let us express the function $Z_{in}(f)$ as a quotient of two polynomials, namely

$$Z_{in} = N(S)/M(S) \tag{5.62}$$

where $N(S) = S^3 + a_1S$, $M(S) = b_2S^2 + b_0$, $S = j\tan(2\pi f t_0)$ and a_1, b_0 and b_2 are real coefficients. From (5.62), it follows that at any frequency f, the response $Z_{in}(f)$ achieves its minimal value equal to zero when the polynomial $N(S)$ takes the zero value at this frequency. Similarly, from (5.62) and Figure 5.30, the following condition for current resonance results

$$M(S) + j2\pi f C N(S) = 0 \tag{5.63}$$

Hence, let us assume that the values of frequency f_1 and capacitance C are given first. Then the designing of the filter consists of evaluating the suitable values

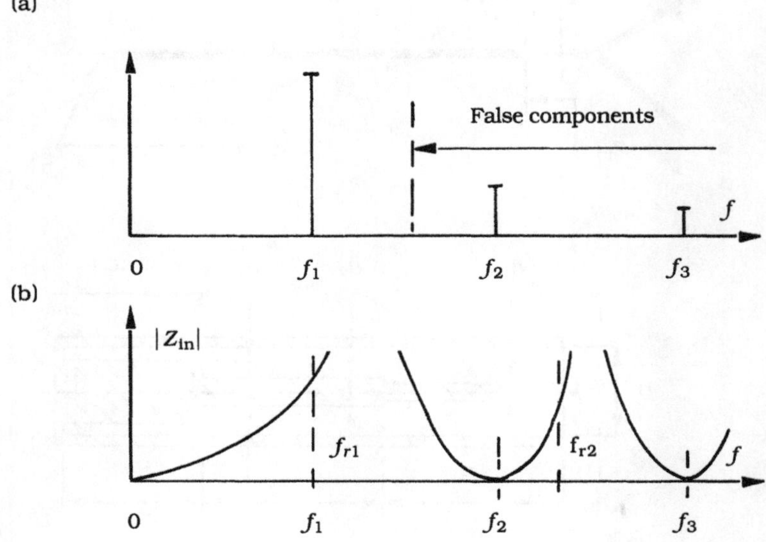

Figure 5.31 The frequency responses: (a) magnitude spectrum of the signal being filtered and (b) modulus of the driving point impedance.

of t_0, Z_{01}, Z_{02} and Z_{03}. In our sample algorithm here, we assume that $t_0 = 1/(10f_1)$ and $f_{r2} = 2.4f_1$, where f_{r2} is the frequency of the second current resonance of the entire filter (see Figure 5.31). Under this assumption, the signals of frequencies $f_2 = 2f_1, f_3 = 3f_1$, and $f_5 = 5f_1$ are fully filtered (i.e., rejected towards the source). It is easy to prove that conditions $Z_{in}(f_2) = 0$, $Z_{in}(f_3) = 0$ and $Z_{in}(f_5) = 0$ are satisfied when $a_1 = \tan^2(2\pi f_2 t_0) = 9.472136$. The coefficients b_0 and b_2 of polynomial $M(S)$ are calculated from (5.63) which should be satisfied at frequencies $f_{r1} = f_1$ and $f_{r2} = 2.4f_1$. Thus:

$$-0.527864b_2 + b_0 = 2\pi f_1 C\, 6.49839$$

$$-252.636b_2 + b_0 = 2\pi f_1 C(-3864.99)$$

By solving the above linear equations, we find the coefficients b_0 and b_2. The next step in the design process is to calculate impedances Z_{01}, Z_{02} and Z_{03} by using the following formulas:

$$Z_{0i} = 1/Q_i \quad \text{for } i = 1, 2 \text{ and } 3 \tag{5.64}$$

where

$$Q_3 = (b_0 + b_2)/(1 + a_1)$$

$$Q_2 = (b_0 + Q_3)/(a_1 - b_0/Q_3)$$

$$Q_1 = b_0 Q_2/Q_3$$

The geometrical lengths of the particular sections (see Figure 5.30) are

$$l_i = 3\ 10^8/(10 f_1 \sqrt{\epsilon_{efi}})\ \text{m} \tag{5.65}$$

where $i = 1, 2$ and 3, and ϵ_{efi} is the effective permittivity of the ith section [5,17,20].

The design algorithm presented above has been implemented in the computer program HF. The initial input data for this program are:

f_1 = frequency of the signal being filtered in Hz,
C = value of the lumped capacitance in F.

We obtain values of delay time t_0 and characteristic impedances Z_{01}, Z_{02} and Z_{03} as output results. Moreover, the computer program HF also makes possible calculation of some geometrical dimensions of the filter composed of coaxial line segments. For this purpose, we must run the program HF again, by using the command CONTINUE, and then introduce the suitable structural data requested by the computer. The meaning of these data is given in Figure 5.30(b).

In most practical cases, the largest step discontinuity occurs between the first and second section. In program HF, this discontinuity is calculated by using (5.50), and compensated at frequency f_2. Finally, the corrected length of the first section is

$$l_{1c} = \frac{3\ 10^8}{4\pi f_1 \sqrt{\epsilon_r}} \arctan\left(\frac{q}{1 + 4\pi f_1 C_f q Z_{01}}\right) \text{ m}$$

where $q = \tan(2\pi/5)$, ϵ_r is the relative permittivity of the dielectric substrate and C_f is the equivalent capacitance of the step discontinuity.

The flow diagram of computer program HF is shown in Figure 5.32.

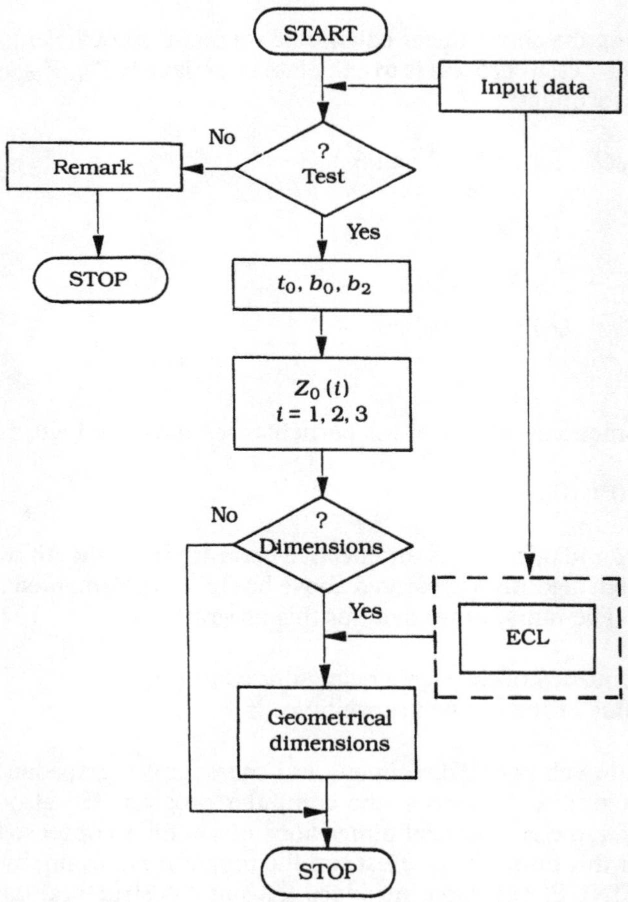

Figure 5.32 The flow diagram of the HF computer program.

Check Calculations

The computer program HF, presented below, has been tested for the following data: $f_1 = 10^9$ Hz and $C = 5\ 10^{-13}$ F. We have assumed that the filter is constructed of coaxial line segments with $D = 7\ 10^{-3}$ m and $\epsilon_r = 1$ (see Figure 5.30(b)).

The values we obtained for characteristic impedances and geometrical dimensions of the coaxial line sections when creating the designed filter are summarized in Table 5.10.

Table 5.10
Characteristic Impedances and Geometrical Dimensions of the Multiharmonic Bandstop Filter Shown in Figure 5.30

	$t_0 = 10^{-10}$ s, $C_f = 1.1\ 10^{-13}$ F		
i	Z_{0i}, ohm	d_i, m	l_i, m
1	11.85	$5.74\ 10^{-3}$	$29.64\ 10^{-3}$
2	51.29	$2.97\ 10^{-3}$	$30.00\ 10^{-3}$
3	53.12	$2.88\ 10^{-3}$	$30.00\ 10^{-3}$

Listing of Program HF

```
6600 REM HF
6602 CLS
6604 PRINT "MULTIHARMONIC BAND-STOP FILTER"
6606 PRINT
6608 PRINT "Data:"
6610 PRINT
6612 LET PI=3.141593
6614 INPUT "fr[Hz]=",FR
6616 INPUT "C[F]=",C
6618 PRINT
6620 IF FR <= 0 THEN  PRINT "Error: fr <= 0": PRINT : GOTO 6614
6622 IF C <= 0 THEN  PRINT "Error: C <= 0": PRINT : GOTO 6616
6624 IF FR*C>.0002 THEN  GOTO 6632
6626 PRINT "Increase the capacitance C"
6628 PRINT
6630 GOTO 6616
6632 IF FR*C<.001 THEN  GOTO 6640
6634 PRINT "Reduce the capacitance C"
6636 PRINT
6638 GOTO 6616
6640 DIM D(3)
6642 DIM L(3)
6644 DIM W(3)
6646 LET T=1/(10*FR)
6648 LET K1=2* PI *FR*C*6.49839
6650 LET K2=4.8* PI *FR*C*-3864.99
6652 LET A1=9.472136
6654 LET B2=(K1-K2)/(252.636-.527864)
```

```
6656 LET B0=K2+B2*252.636
6658 LET W(3)=(B0+B2)/(1+A1)
6660 LET W(2)=(B0+W(3))/(A1-B0/W(3))
6662 LET W(1)=W(2)*B0/W(3)
6664 PRINT "Results, see Fig. 5.30(a)"
6666 PRINT
6668 PRINT "to[s]=";T
6670 PRINT
6672 FOR I=1 TO 3
6674 PRINT "Zo(";I;")[ohm]=";1/W(I)
6676 NEXT I
6678 GOSUB 6734
6680 PRINT
6682 PRINT "Enetr the outer diameter D, see Fig. 5.30(b)"
6684 PRINT
6686 INPUT "D[m]=",D
6688 PRINT
6690 PRINT "Geometrical dimensions in meters:"
6692 PRINT
6694 FOR I=1 TO 3
6696 LET D(I)=D/ EXP (1/(W(I)*60))
6698 PRINT "d(";I;")=";D(I)
6700 NEXT I
6702 PRINT
6704 LET CD= PI *D/(10^12)*(13+.2*D/D(2))*((D(1)-D(2))/(D-D(2))
)^2.65
6706 LET O=2* PI *FR*T
6708 LET O1= ATN ( TAN (O)/(1+2* PI *FR*CD* TAN (O)/W(1)))
6710 LET T1=O1/(2* PI *FR)
6712 LET L(1)=3*10^8*T1
6714 LET L(2)=3*10^8*T
6716 LET L(3)=3*10^8*T
6718 FOR I=1 TO 3
6720 PRINT "l(";I;")=";L(I)
6722 NEXT I
6724 REM Chain subroutine
6726 PRINT
6728 INPUT"Enter 'CONT':",N$
6730 IF N$="CONT" THEN CLS:CHAIN "MENU",10
6732 GOTO 6726
6734 REM Control subroutine
6736 PRINT
6738 INPUT "Enter 'CONT':",N$
6740 IF N$="CONT" THEN CLS:RETURN
6742 GOTO 6736
```

5.10 LUMPED ELEMENT RESISTANCE ATTENUATORS

The topology of T and π resistance attenuators, under consideration in this section, is shown in Figure 5.33. Attenuators of this type are mainly used to reduce power in a controllable known degree, to match the terminating resistances over a very wide frequency range, and also to reduce amplifier gain (as well as power when tuning them). We should remember, however, that these requirements can be satisfied only if the assumed power attenuation $L = 10\log(P_{in}/P_{out})$ dB is not smaller than the threshold attenuation, L_{min}, given by:

$$L_{min} = 10\log(\sqrt{r} + \sqrt{r-1})^2, \text{ dB} \tag{5.66}$$

where $r = |Z_{02}/Z_{01}|^{\pm 1} \geq 1$ is the ratio of resistances that are to be matched. When condition $L \geqslant L_{min}$ is satisfied, the corresponding resistances of the attenuator being designed may be calculated by using the following formulas.

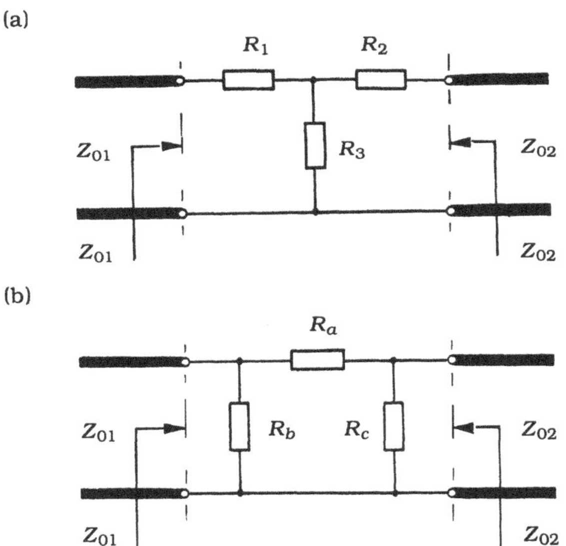

Figure 5.33 Resistance attenuators: (a) of T type and (b) of π type.

For the T structure attenuator (see Figure 5.33(a)).

$$\begin{aligned} R_3 &= 2\sqrt{NZ_{01}Z_{02}}/(N-1) \\ R_1 &= Z_{01}(N+1)/(N-1) - R_3 \\ R_2 &= Z_{02}(N+1)/(N-1) - R_3 \end{aligned} \tag{5.67}$$

where $N = 10^{(L/10)}$, and L is the required attenuation in dB.

For the π structure attenuator (see Figure 5.33(b))

$$R_a = (N - 1)\sqrt{Z_{01}Z_{02}}/(2\sqrt{N})$$
$$R_b = Z_{01}(N - 1)R_a/[(N + 1)R_a - Z_{01}(N - 1)] \quad (5.68)$$
$$R_c = Z_{02}(N - 1)R_a/[(N + 1)R_a - Z_{02}(N - 1)]$$

Also in this case $N = 10^{(L/10)}$, where L is the required attenuation expressed in dB.

From (5.67) and (5.68), it follows that for $L = L_{min}$, the T and π structure attenuators reduce to Γ attenuators, as shown in Figure 5.34.

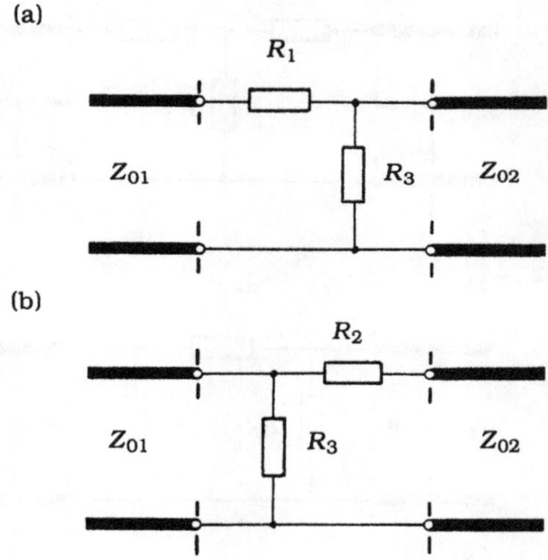

Figure 5.34 Resistance attenuators: (a) of Γ type for $Z_{01} > Z_{02}$ and (b) of Γ type for $Z_{01} < Z_{02}$.

The main advantage of these design formulas lies in their simple mathematical form. This form led to their use in the computer program RA presented below. This program allows us to calculate the resistances of T or π attenuators when the acceptable values of Z_{01}, Z_{02} and L are given. Moreover, we can design the T and π attenuators with minimum value of attenuation equal to L_{min}. For this design, we must introduce into the computer memory the text variable L_m instead of the value of attenuation L.

The flow diagram of the computer program RA is shown in Figure 5.35.

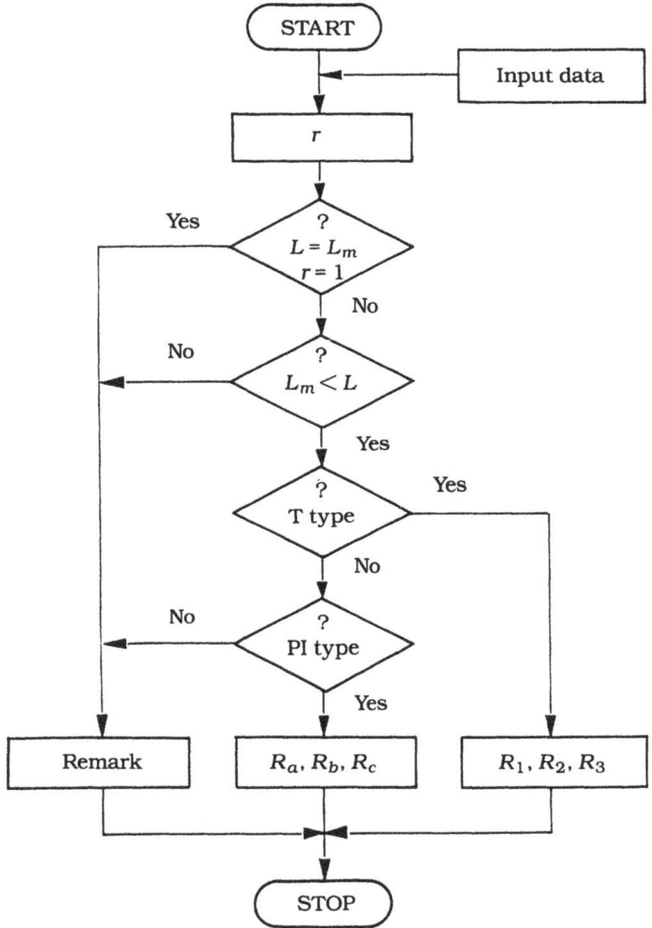

Figure 5.35 The flow diagram of the RA computer program.

Example of Calculations

The computer program RA may be tested by using the computational data given in Tables 5.11 and 5.12.

Table 5.11
Some Results for *T* Type Resistance Attenuators

	T Attenuator (See Figure 5.33(a))				
L, dB	Z_{01}*, ohm*	Z_{02}*, ohm*	R_1*, ohm*	R_2*, ohm*	R_3*, ohm*
3	50.00	50.00	8.55	8.55	141.92
3	50.00	60.00	—	—	—
6	50.00	60.00	10.22	26.93	73.32
L_m	50.00	75.00	0.00	43.30	86.60
6	50.00	75.00	1.57	43.34	81.97

Table 5.12
Some Results for π Type Resistance Attenuators

	π Attenuator (See Figure 5.33(b))				
L, dB	Z_{01}*, ohm*	Z_{02}*, ohm*	R_a*, ohm*	R_b*, ohm*	R_c*, ohm*
3	50.00	50.00	17.61	292.40	292.40
3	50.00	60.00	—	—	—
6	50.00	60.00	40.91	111.38	293.37
L_m	50.00	75.00	43.30	86.60	$> 10^{10}$
6	50.00	75.00	45.74	86.51	2386.20

Note: $L_m = 5.719$ dB.

Listing of Program RA

```
6800 REM RA
6802 CLS
6804 PRINT "RESISTANCE ATTENUATORS"
6806 PRINT
6808 PRINT "Data:"
6810 PRINT
6812 INPUT "Zo1[ohm]=",ZO1
6814 INPUT "Zo2[ohm]=",ZO2
6816 INPUT"L[dB]=",L$
6818 IF ZO1 <= 0 THEN  PRINT "Error:Zo1 <= 0": GOTO 6810
6820 IF ZO2 <= 0 THEN  PRINT "Error:Zo2 <= 0": GOTO 6810
6822 PRINT
6824 IF ZO1 >= ZO2 THEN  LET R=ZO1/ZO2: GOTO 6828
6826 LET R=ZO2/ZO1
6828 IF L$="Lm" AND R=1 THEN  PRINT "The terminating resistance
s can be connected directly !": GOTO 6906
6830 IF L$="Lm" AND R>1 THEN  LET N=2*R+2* SQR (R*R-R)-1: GOTO
6860
6832 LET L= VAL(L$)
6834 LET LM=10/ LOG(10)* LOG(( SQR(R)+ SQR (R-1))^2)
```

```
6836 IF L=LM AND R>1 THEN  GOTO 6860
6838 IF L=LM AND R=1 THEN  PRINT "The terminating resistances c
an be connected directly !": GOTO 6906
6840 IF L >= LM THEN  GOTO 6846
6842 PRINT "L<Lm !": GOTO 6906
6844 PRINT
6846 INPUT "Enter the name ( T or PI )  of the attenuator to be
 designed:",A$
6848 PRINT
6850 LET N=10^(L/10)
6852 IF A$="T" THEN  GOTO 6860
6854 IF A$="PI" THEN  GOTO 6888
6856 PRINT
6858 PRINT "Wrong commamd !": GOTO 6844
6860 LET R3=2* SQR (N*Z01*Z02)/(N-1)
6862 LET R1=Z01*(N+1)/(N-1)-R3
6864 LET R2=Z02*(N+1)/(N-1)-R3
6866 IF R1<.000001 THEN  LET R1=0
6868 IF R2<.000001 THEN  LET R2=0
6870 PRINT
6872 PRINT "Results, see Fig. 5.33(a)"
6874 PRINT
6876 PRINT "R1[ohm]=";R1
6878 PRINT "R2[ohm]=";R2
6880 PRINT "R3[ohm]=";R3
6882 PRINT
6884 IF L$="Lm" THEN  PRINT "Lm[dB]=";10/ LOG(10)* LOG((SQR(R)+
 SQR (R-1))^2)
6886 GOTO 6906
6888 LET RA=(N-1)/2* SQR (Z01*Z02/N)
6890 LET RB=Z01*(N-1)*RA/((N+1)*RA-Z01*(N-1))
6892 LET RC=Z02*(N-1)*RA/((N+1)*RA-Z02*(N-1))
6894 PRINT
6896 PRINT "Results, see Fig. 5.33(b)"
6898 PRINT
6900 PRINT "Ra[ohm]=";RA
6902 PRINT "Rb[ohm]=";RB
6904 PRINT "Rc[ohm]=";RC
6906 REM Chain subroutine
6908 PRINT
6910 INPUT"Enter 'CONT':",N$
6912 IF N$="CONT" THEN CLS:CHAIN "MENU",10
6914 GOTO 6908
```

5.11 NARROW-BAND IMPEDANCE MATCHING CIRCUITS WITH LUMPED AND DISTRIBUTED PARAMETERS

From a practical standpoint, impedance matching is designing an interconnecting circuit in such a way that the terminating impedance is transformed exactly into a desired impedance at a single frequency, or is transformed approximately over a band of frequencies. In this section, we consider impedance matching at a single frequency, first with two- and three-element circuits composed of lossless lumped inductors and capacitors, and then with lossless one- and two-section TEM transmission lines. Electrical diagrams of these circuits are shown in Figures 5.36, 5.37, 5.38 and 5.39, respectively.

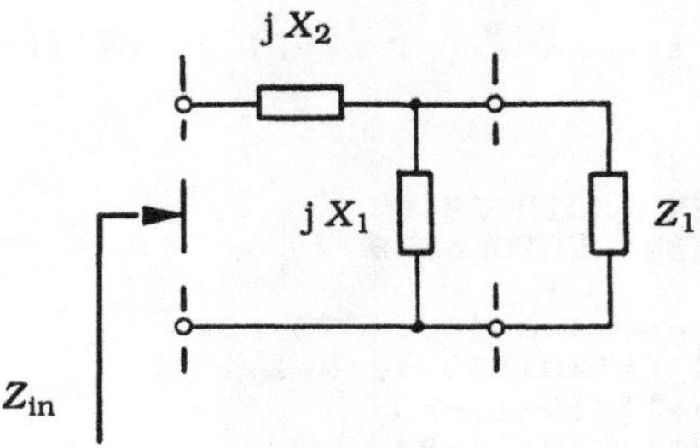

Figure 5.36 The two-element impedance matching circuit.

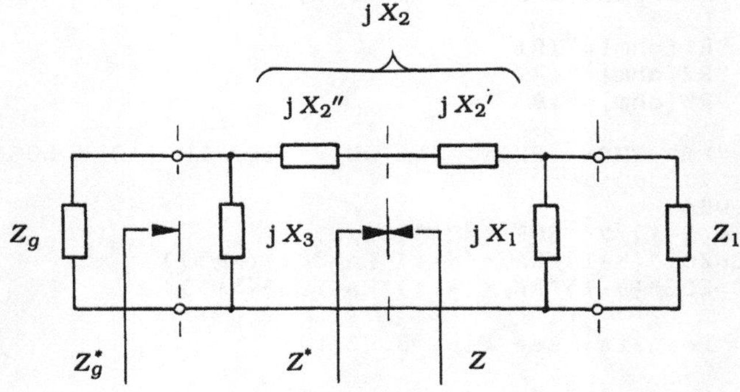

Figure 5.37 The three-element impedance matching circuit.

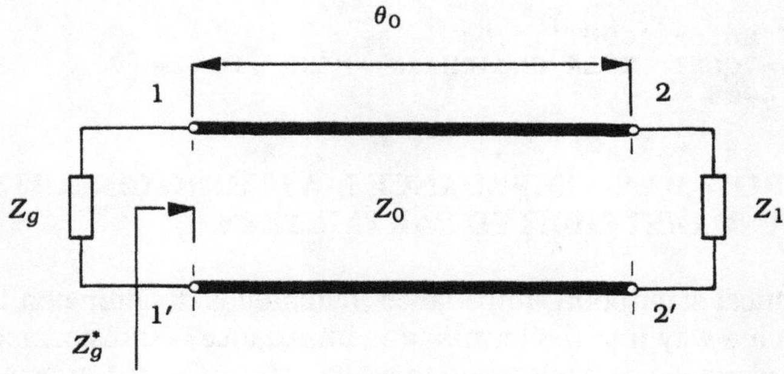

Figure 5.38 The one-section transmission line impedance matching circuit.

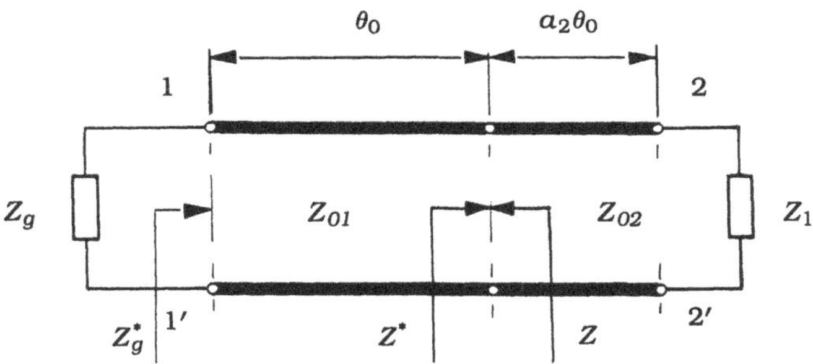

Figure 5.39 The two-section transmission line impedance matching circuit.

In considering the basic relationships for these circuits, we assume a conjugate matching associated with maximum power transfer from the source into the load [20].

At the beginning, let us consider the simplest matching circuit (Γ section) shown in Figure 5.36. The load impedance $Z_l = R_l + jX_l$ is transformed accurately to a desired impedance $Z_{in} = R_{in} + jX_{in}$ if

$$X_1 = \frac{R_{in}X_l \pm \sqrt{R_{in}^2 X_l^2 + R_{in}(R_l - R_{in})(R_l^2 + X_l^2)}}{R_l - R_{in}}$$
$$X_2 = X_{in} - X_1 \frac{R_l^2 + X_l(X_l + X_1)}{R_l^2 + (X_l + X_1)^2} \tag{5.69}$$

From (5.69), we see that reactance X_1 can take two different values, the first negative and the second positive. This means that the circuit being designed can be realized in *LC* and *CL* configurations.

The formulas in (5.69) are also helpful for designing the π matching circuit presented in Figure 5.37, because this circuit can be treated as composed of two Γ sections. In this case, values of reactances X_1, X_2', X_3 and X_2'' are calculated successively by using (5.69) and the auxiliary impedance $Z = R + j0$ with $R < \min(R_g, R_l)$.

We have found that when R is decreased with respect to $\min(R_g, R_l)$, the matching band narrows.

The one-section transmission line (see Figure 5.38) is the simplest matching circuit with distributed parameters. In [22], the conclusion is that, in many applications, this type of transformer is quite adequate. Its design is very simple, and consists of calculating values of characteristic impedance Z_0 and electrical length θ_0 from the following relations:

$$Z_0 = \sqrt{[R_l(R_g^2 + X_g^2) - R_g(R_l^2 + X_l^2)]/(R_g - R_l)}$$
$$\theta_0 = \arctan[Z_0(R_l - R_g)/(X_g R_l - X_l R_g)] \quad (5.70)$$

where $Z_g = R_g + jX_g$ and $Z_l = R_l + jX_l$ are impedances that are to be matched. From (5.70), we see that not all impedance combinations can be matched in this way. The quantity under the square root must always be positive. In practice, however, applications of this matching circuit type are limited mainly by the relatively narrow interval of characteristic impedances Z_0, $Z_{0\,min} \leqslant Z_0 \leqslant Z_{0\,max}$, that can be realized without difficulty. These constraints can be expressed in terms of impedances Z_g, Z_l, $Z_{0\,min}$ and $Z_{0\,max}$ as follows:

(a) $Z_0 \geqslant Z_{0\,min}$, if

$$X_l^2 + (R_l - l_1)^2 \geqslant p_1^2 \quad \text{for } R_l \geqslant R_g \quad (5.71)$$
$$X_l^2 + (R_l - l_1)^2 \leqslant p_1^2 \quad \text{for } R_l \leqslant R_g$$

where

$$l_1 = (X_g^2 + R_g^2 + Z_{0\,min}^2)/(2R_g)$$
$$p_1 = \sqrt{l_1^2 - Z_{0\,min}^2}$$

(b)

$Z_0 \leqslant Z_{0\,max}$, if (5.72)

$$X_l^2 + (R_l - l_1')^2 \leqslant p_1'^2 \quad \text{for } R_l \geqslant R_g$$
$$X_l^2 + (R_l - l_1')^2 \geqslant p_1'^2 \quad \text{for } R_l \leqslant R_g$$

where

$$l_1' = (X_g^2 + R_g^2 + Z_{0\,max}^2)/(2R_g)$$
$$p_1' = \sqrt{l_1'^2 - Z_{0\,max}^2}$$

The graphical interpretation of (5.71) and (5.72) is given in Figure 5.40. The above mathematical relations allow us to analyze the two-section matching circuit the electrical scheme of which is shown in Figure 5.39. In this case, the given impedances Z_g and Z_l can be matched when auxiliary impedances $Z = R + jX$ and $Z^* = R - jX$ lie in the permissible area A_c (see Figure 5.41). In general, this area is an intersection of the corresponding permissible areas evaluated for impedances Z_g and Z_l in the manner described above. Here, we note that impedances Z and Z^* determine the values of Z_{01}, Z_{02}, θ_0 and a_2, the meaning of which is given in Figure 5.39. In other words, the circuit under consideration can be optimized with respect to its fractional bandwidth or total electrical length by the suitable choice of impedance Z. At lower microwave frequencies, the circuits optimized with respect to the total electrical length $\theta_{t0} = \theta_0(1 + a_2)$ are of particular interest. Hence, let us formulate the optimization problem as:

$$\min_{Z \subset A_c} \quad \theta_{t0}(Z) \tag{5.73}$$

where $\subset$ means that the impedance Z lies within the area A_c.

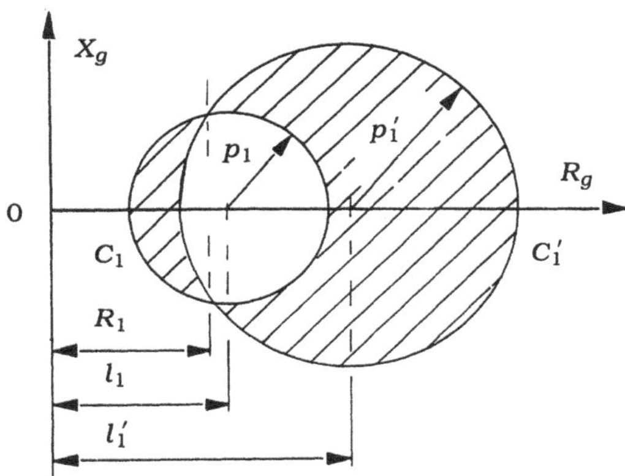

Figure 5.40 The area of complex impedance Z_g which can be matched to the complex impedance Z_l at constraints $Z_0 \geqslant Z_{0\ min}$ and $Z_0 \leqslant Z_{0\ max}$.

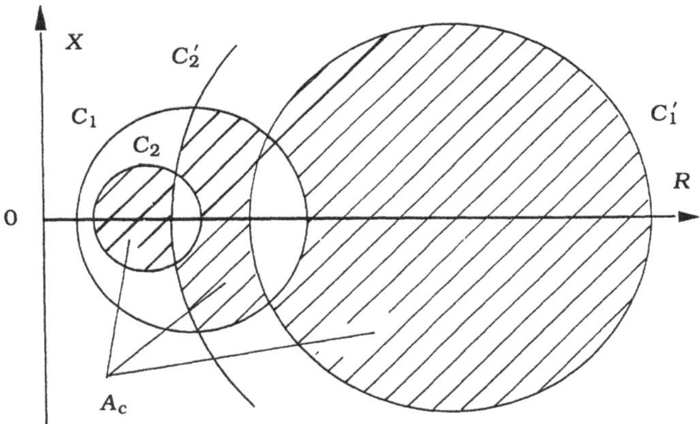

Figure 5.41 The area of complex impedances Z and Z* which can be used to design the matching circuit shown in Figure 5.39.

The numerical experiment confirms that the circuit being analyzed is optimal in the sense of (5.73) when impedances Z_{01} and Z_{02} (see Figure 5.39) take extreme permissible values (i.e., $Z_{0\ min}$ and $Z_{0\ max}$) or vice versa. This means that the optimum value of impedance Z (as well as Z^*) lies on the border of area A_c, and is determined by the intersection of circles C_1 and C_2' or C_2 and C_1' as shown in Figure 5.41. Hence, the optimum value of impedance Z can be evaluated analytically from relationships such as (5.71) and (5.72). We have taken advantage of this favorable opportunity in the computer program NMC, presented below. In this program, the electromagnetic field discontinuities due to step changes in the geometry of the particular line sections are also taken into account. As a matter of fact, the enlarged circuit shown in Figure 5.42 is designed. The step reactances are compensated in the manner described in [23 and 24]. For zero values of step reactance elements C_{e1}, C_{e2}, C_{s1}, C_{s2}, L_{s1} and L_{s2} this circuit reduces to its fundamental form, presented in Figure 5.39.

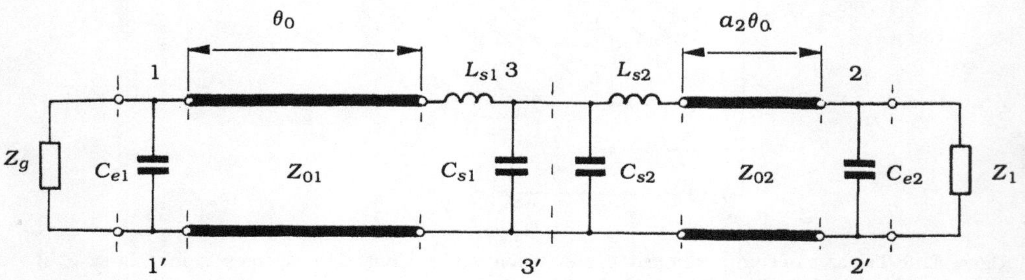

Figure 5.42 The extended electrical scheme of the two-section transmission line impedance matching circuit.

The manner of using the computer program NMC for designing the matching circuits discussed here is illustrated by some check calculations presented below.

Check Calculations

Example 1

Input Data:

MC/I—brief name of the matching circuits shown in Figure 5.36

$Z_{in} = 30 - j10$ Ω

$Z_l = 50 + j20$ Ω

Results:

$X_1 = -42.45$ Ω

$X_{2'} = 18.98$ Ω

or

$X_1 = 102.45$ Ω
$X_2 = -38.98$ Ω

Example 2

Input Data:

MC/II—brief name of the matching circuit shown in Figure 5.37

$Z_g = 30 + j10$ Ω
$Z_l = 50 + j0$ Ω
$R = 20$ Ω

Results:

$X_1 = 40.82$ Ω
$X_2 = -40.82$ Ω
$X_3 = 68.99$ Ω

or

$X_1 = -40.82$ Ω
$X_2 = 40.82$ Ω
$X_3 = -28.99$ Ω

Example 3

Input Data:

MC/III—brief name of the matching circuit shown in Figure 5.38

$Z_g = 40 - j70$ Ω
$Z_l = 20 - j30$ Ω
$Z_{0\ min} = 20$ Ω
$Z_{0\ max} = 100$ Ω

Results:

$Z_0 = 62.45$ Ω
$\theta = 1.41$ rad

Example 4

Input Data:

MC/IV—brief name of the matching circuit shown in Figure 5.42

$Z_g = 50 + j20$ Ω
$Z_l = 30 - j10$ Ω

$Z_{0\,min} = 30$ Ω
$Z_{0\,max} = 70$ Ω
$f_0 = 2\ 10^9$ Hz
$C_{e1} = 8.08\ 10^{-14}$ F
$C_{e2} = 2.83\ 10^{-14}$ F
$C_{s1} = 2.26\ 10^{-14}$ F
$C_{s2} = 2.26\ 10^{-14}$ F
$L_{s1} = 2.06\ 10^{-11}$ H
$L_{s2} = 4.59\ 10^{-11}$ H

Results:

$Z_{01} = 29.99$ Ω
$Z_{02} = 73.07$ Ω
$\theta_0 = 0.780$ rad
$a_2 = 0.657$

Listing of Program NMC

```
7000 REM NMC
7002 CLS
7004 PRINT "NARROW BAND IMPEDANCE MATCHING CIRCUITS"
7006 PRINT
7008 PRINT "Data:"
7010 PRINT
7012 INPUT "Enter the name (MC/I, MC/II, MC/III or MC/IV ) of t
he circuit to be designed:",A$
7014 LET PI=3.141593
7016 PRINT
7018 IF A$="MC/I" THEN  GOTO 7032
7020 IF A$="MC/II" THEN  GOTO 7074
7022 IF A$="MC/III" THEN  GOTO 7172
7024 IF A$="MC/IV" THEN  GOTO 7226
7026 PRINT
7028 PRINT "Wrong command !"
7030 GOTO 7010
7032 INPUT ;"Zin[ohm]=",RE:INPUT " +j",XE
7034 INPUT ;"Zl[ohm]=", RL:INPUT " +j",XL
7036 PRINT
7038 IF RE <= 0 THEN  PRINT "Re(Zin) <= 0": GOTO 7032
7040 IF RL <= 0 THEN  PRINT "Re(Zl) <= 0": GOTO 7034
7042 LET P1=RE
7044 LET P2=RL
7046 LET Q1=XE
7048 LET Q2=XL
7050 GOSUB 7212
7052 GOSUB 7218
7054 PRINT "The electrical scheme of MC/I circuit is shown in F
ig. 5.36"
7056 PRINT
```

```
7058 PRINT "X1[ohm]=";X1
7060 PRINT "x2[ohm]=";X2
7062 LET SQ=-SQ
7064 GOSUB 7218
7066 PRINT "or"
7068 PRINT "x1[ohm]=";X1
7070 PRINT "x2[ohm]=";X2
7072 GOTO 7524
7074 INPUT ;"Zg[ohm]=",RG: INPUT " +j",XG
7076 INPUT ;"Zl[ohm]=",RL: INPUT " +j",XL
7078 PRINT
7080 IF RG <= 0 THEN  PRINT "Re(Zg) <= 0": GOTO 7074
7082 IF RL <= 0 THEN  PRINT "Re(Zl) <= 0": GOTO 7076
7084 INPUT "R[ohm]=",R
7086 IF R<RG AND R<RL THEN  GOTO 7096
7088 PRINT
7090 PRINT "Assume R< min(Rg,Rl)"
7092 PRINT
7094 GOTO 7082
7096 LET X=0
7098 PRINT
7100 LET P1=R
7102 LET P2=RL
7104 LET Q1=X
7106 LET Q2=XL
7108 GOSUB 7212
7110 GOSUB 7218
7112 PRINT "The electrical scheme of MC/II circuit is shown in
Fig. 5.37"
7114 PRINT
7116 PRINT "X1[ohm]=";X1
7118 LET XS=X2
7120 LET P2=RG
7122 LET Q1=-X
7124 LET Q2=XG
7126 GOSUB 7212
7128 GOSUB 7218
7130 PRINT "X2[ohm]=";XS+X2
7132 PRINT "X3[ohm]=";X1
7134 PRINT "or"
7136 LET P1=R
7138 LET P2=RL
7140 LET Q1=X
7142 LET Q2=XL
7144 GOSUB 7212
7146 LET SQ=-SQ
7148 GOSUB 7218
7150 PRINT "X1[ohm]=";X1
7152 LET XS=X2
7154 LET P2=RG
7156 LET Q1=-X
7158 LET Q2=XG
7160 GOSUB 7212
7162 LET SQ=-SQ
7164 GOSUB 7218
```

```
7166 PRINT "X2[ohm]=";XS+X2
7168 PRINT "X3[ohm]=";X1
7170 GOTO 7524
7172 INPUT ;"Zg[ohm]=",RG:INPUT " +j",XG
7174 INPUT ;"Zl[ohm]=",RL:INPUT " +j",XL
7176 PRINT
7178 IF RG <= 0 THEN  PRINT "Re(Zg) <= 0": GOTO 7172
7180 IF RL <= 0 THEN  PRINT "Re(Zl) <= 0": GOTO 7174
7182 IF RG=RL AND XG=XL THEN  GOTO 7192
7184 IF RG=RL AND XG=-XL THEN  LET ZZ=RG*RL: GOTO 7190
7186 IF RG=RL THEN  GOTO 7192
7188 LET ZZ=((RG*RG+XG*XG)*RL-(RL*RL+XL*XL)*RG)/(RG-RL)
7190 IF ZZ>0 THEN  GOTO 7196
7192 PRINT "This circuit cannot be realized !"
7194 GOTO 7524
7196 LET ZO= SQR(ZZ)
7198 PRINT "The electrical scheme of MC/III circuit is shown in
 Fig. 5.38"
7200 PRINT
7202 PRINT "Zo[ohm]=";ZO
7204 LET O= ATN (ZO*(RL-RG)/(XG*RL-XL*RG))
7206 IF O<0 THEN  LET O= PI +O
7208 PRINT "0[rad]=";O
7210 GOTO 7524
7212 REM Subroutine 7212
7214 LET SQ= SQR (P1*P1*Q2*Q2+P1*(P2-P1)*(P2*P2+Q2*Q2))
7216 RETURN
7218 REM Subroutine 7218
7220 LET X1=(P1*Q2+SQ)/(P2-P1)
7222 LET X2=Q1-X1*(Q2*(Q2+X1)+P2*P2)/((Q2+X1)*(Q2+X1)+P2*P2)
7224 RETURN
7226 INPUT ;"Zg[ohm]=",RG:INPUT " +j",XG
7228 INPUT ;"Zl[ohm]=",RL:INPUT " +j",XL
7230 PRINT
7232 IF RG <= 0 THEN  PRINT "Re(Zg) <= 0": GOTO 7226
7234 IF RL <= 0 THEN  PRINT "Re(Zl) <= 0": GOTO 7228
7236 INPUT "Zomin[ohm]=",ZN
7238 INPUT "Zomax[ohm]=",ZX
7240 IF ZN <= 0 THEN  PRINT "Error:Zomin <= 0": GOTO 7236
7242 IF ZX <= 0 THEN  PRINT "Error:Zomax <= 0": GOTO 7238
7244 PRINT
7246 PRINT "Please wait !"
7248 PRINT
7250 IF RG=RL AND XG=-XL THEN  PRINT "Connect directly !": GOTO
 7524
7252 DIM Z(4)
7254 DIM R(4)
7256 LET IT=1
7258 LET R1=RG
7260 LET X1=XG
7262 LET R2=RL
7264 LET X2=XL
7266 LET L1=(X1*X1+R1*R1+ZN*ZN)/(2*R1)
7268 LET L2=(X2*X2+R2*R2+ZN*ZN)/(2*R2)
7270 LET P1= SQR (L1*L1-ZN*ZN)
```

```
7272 LET P2= SQR (L2*L2-ZN*ZN)
7274 LET LL1=(X1*X1+R1*R1+ZX*ZX)/(2*R1)
7276 LET LL2=(X2*X2+R2*R2+ZX*ZX)/(2*R2)
7278 LET PP1= SQR (LL1*LL1-ZX*ZX)
7280 LET PP2= SQR (LL2*LL2-ZX*ZX)
7282 PRINT
7284 LET FL=0
7286 LET R=(P1*P1-PP2*PP2-L1*L1+LL2*LL2)/(2*(LL2-L1))
7288 IF  ABS (R-L1)>P1 THEN  LET FL=1: GOTO 7298
7290 LET R(1)=R
7292 LET R(2)=R
7294 LET X(1)= SQR (ABS (P1*P1-(R-L1)*(R-L1)))
7296 LET X(2)=-X(1)
7298 LET R=(P2*P2-PP1*PP1-L2*L2+LL1*LL1)/(2*(LL1-L2))
7300 IF  ABS (R-L2)>P2 AND FL=1 THEN  PRINT "The solution canno
t be found !": GOTO 7524
7302 IF  ABS (R-L2)>P2 AND FL=0 THEN  LET R(3)=R(1): LET R(4)=R
(2): LET X(3)=X(1): LET X(4)=X(2): GOTO 7322
7304 LET R(3)=R
7306 LET R(4)=R
7308 LET X(3)= SQR (ABS (P2*P2-(R-L2)*(R-L2)))
7310 LET X(4)=-X(3)
7312 IF FL=0 THEN  GOTO 7322
7314 LET R(1)=RR
7316 LET R(2)=RR
7318 LET X(1)=X(3)
7320 LET X(2)=X(4)
7322 LET OT=2* PI
7324 FOR I=1 TO 4
7326 LET RA=R(I)
7328 LET XA=X(I)
7330 LET RB=R1
7332 LET XB=X1
7334 GOSUB 7492
7336 LET Z1=ZO
7338 IF OO<0 THEN  LET O1=OO+ PI : GOTO 7342
7340 LET O1=OO
7342 LET XA=-X(I)
7344 LET RB=R2
7346 LET XB=X2
7348 GOSUB 7492
7350 LET Z2=ZO
7352 IF OO<0 THEN  LET O2=OO+ PI : GOTO 7356
7354 LET O2=OO
7356 LET OC=O1+O2
7358 IF OC<OT THEN  LET ZO1=Z1: LET ZO2=Z2: LET OO1=O1: LET OO2
=O2: LET R3=R(I): LET X3=X(I): LET OT=OC
7360 NEXT I
7362 IF IT >= 2 THEN  GOTO 7432
7364 PRINT "Initial results"
7366 PRINT
7368 PRINT "Zo1[ohm]=";ZO1
7370 PRINT "Zo2[ohm]=";ZO2
7372 PRINT "Oo[rad]=";OO1
7374 PRINT "a2=";OO2/OO1
```

```
7376 PRINT
7378 PRINT "Enter the parasitic elements, see Fig. 5.42"
7380 PRINT
7382 INPUT "Ce1[F]=",CE1
7384 INPUT "Ce2[F]=",CE2
7386 INPUT "Cs1[F]=",CS1
7388 INPUT "Cs2[F]=",CS2
7390 INPUT "Ls1[H]=",LS1
7392 INPUT "Ls2[H]=",LS2
7394 PRINT
7396 PRINT "Enter the center band frequency"
7398 PRINT
7400 INPUT "fo[Hz]=",FO
7402 PRINT
7404 LET CP=CE1
7406 LET RP=RG
7408 LET XP=XG
7410 GOSUB 7512
7412 LET R1=RE
7414 LET X1=IM
7416 LET CP=CE2
7418 LET RP=RL
7420 LET XP=XL
7422 GOSUB 7512
7424 LET R2=RE
7426 LET X2=IM
7428 LET IT=IT+1
7430 GOTO 7266
7432 LET RP=R3
7434 LET XP=X3
7436 LET CP=CS1
7438 GOSUB 7512
7440 LET RB=RE
7442 LET XB=IM+2* PI *FO*LS1
7444 LET RA=R1
7446 LET XA=X1
7448 GOSUB 7492
7450 LET ZC1=ZO
7452 IF OO<0 THEN  LET OC1=OO+ PI : GOTO 7456
7454 LET OC1=OO
7456 LET XP=-X3
7458 LET CP=CS2
7460 GOSUB 7512
7462 LET RB=RE
7464 LET XB=IM+2* PI *FO*LS2
7466 LET RA=R2
7468 LET XA=X2
7470 GOSUB 7492
7472 LET ZC2=ZO
7474 IF OO<0 THEN  LET OC2=OO+ PI : GOTO 7478
7476 LET OC2=OO
7478 PRINT "Final results, see Fig. 5.42"
7480 PRINT
7482 PRINT "Zo1[ohm]=";ZC1
7484 PRINT "Zo2[ohm]=";ZC2
```

```
7486 PRINT "0o[rad]=";OC1
7488 PRINT "a2=";OC2/OC1
7490 GOTO 7524
7492 REM Subroutine 7492
7494 IF RA=RB AND XA=-XB THEN  LET OO=0: LET ZO=RA: RETURN
7496 LET ZO= SQR (ABS ((RB*(RA*RA+XA*XA)-RA*(RB*RB+XB*XB))/(RA-
RB)))
7498 LET ML=ZO*(RA-RB)
7500 LET MM=RA*XB-RB*XA
7502 IF ML=0 AND MM=0 THEN  LET OO=0: RETURN
7504 IF ML>0 AND MM=0 THEN  LET OO= PI /2: RETURN
7506 IF ML<0 AND MM=0 THEN  LET OO=- PI /2: RETURN
7508 LET OO= ATN (ML/MM)
7510 RETURN
7512 REM Subroutine 7512
7514 LET A1=1-2* PI *FO*CP*XP
7516 LET A2=2* PI *FO*CP*RP
7518 LET RE=(RP*A1+XP*A2)/(A1*A1+A2*A2)
7520 LET IM=(XP*A1-RP*A2)/(A1*A1+A2*A2)
7522 RETURN
7524 REM Chain subroutine
7526 PRINT
7528 INPUT"Enter 'CONT':",N$
7530 IF N$="CONT" THEN CLS:CHAIN "MENU",10
7532 GOTO 7526
```

REFERENCES

[1] Collin, R.E., "Theory and Design of Wide-Band Multisection Quarter-Wave Transformers," *IRE Proc.*, Vol. 43, No. 2, February 1955, pp. 179–185.

[2] Cohn, S.B., "Optimum Design of Stepped Transmission-Line Transformers," *IRE Trans., Microwave Theory and Techniques,* MTT-3, No. 4, April 1955, pp. 16–21.

[3] Riblet, H.J., "General Synthesis of Quarter-Wave Impedance Transformers," *IRE Trans., Microwave Theory and Techniques,* MTT-5, No. 1, January 1957, pp. 34–43.

[4] Young, L., "The Quarter-Wave Transformer Prototype Circuit," *IRE Trans., Microwave Theory and Techniques,* MTT-8, No. 9, September 1960, pp. 483–489.

[5] Matthaei, G.L., L. Young and E.M.T. Jones, *Microwave Filters, Impedance Matching Networks and Coupling Structures,* Norwood, MA: Artech House, 1980.

[6] Feldstein, A.L. and L.R. Javich, *Synthesis of Microwave Two-Ports and Four-Ports* (in Russian). Moscow: Sviaz, 1971.

[7] Matthaei, G.L., "Short-Step Chebyschev Impedance Transformers," *IEEE Trans., Microwave Theory and Techniques,* MTT-14, No. 8, August 1966, pp. 372–383.

[8] Rosłoniec, S., "A New Approach to the Synthesis of Microwave Impedance Transformers," *International Journal of Electronics,* Vol. 64, No. 6, June 1988, pp. 947–954.

[9] Rosłoniec, S., "Design of *R*-transformers on the Basis of the Insertion Loss Function," *International Journal of Electronics,* Vol. 64, No. 3, March 1988, pp. 385–394.

[10] Howe, H., Jr., *Stripline Circuit Design,* Norwood, MA: Artech House, 1974.

[11] Edwards, T.C., *Foundations for Microstrip Circuit Design,* New York: John Wiley & Sons, 1982.

[12] Levy, R., "General Synthesis of Asymmetric Multi-Element Transmission-line Directional Cou-

plers," *IEEE Trans., Microwave Theory and Techniques,* MTT-11, No. 7, July 1963, pp. 226–237.

[13] Arndt, F., "Tables for Asymmetric Chebyschev High-Pass TEM-mode Directional Couplers," *IEEE Trans., Microwave Theory and Techniques,* MTT-18, No. 9, September 1970, pp. 633–638.

[14] Wilkinson, E.J., "An *n*-way Hybrid Power Divider," *IRE Trans., Microwave Theory and Techniques,* MTT-8, No. 1, January 1960, pp. 116–118.

[15] Parad, L. and R. Moynihan, "Split-Tee Power Divider," *IEEE Trans., Microwave Theory and Techniques,* MTT-13, No. 1, January 1965, pp. 91–95.

[16] Cohn, S. B., "A Class of Broad-Band Three-Port TEM-Mode Hybrids," *IEEE Trans., Microwave Theory and Techniques,* MTT-16, No. 2, February 1968, pp. 110–116.

[17] Gupta, K.C., R. Garg and R. Chadha, *Computer-Aided Design of Microwave Circuits,* Norwood, MA: Artech House, 1981.

[18] Cohn, S.B., "Parallel Coupled Transmission Line Resonator Filters," *IRE Trans., Microwave Theory and Techniques,* MTT-6, No. 4, April 1958, pp. 223–231.

[19] Schiffman, B.M. and G.L. Matthaei, "Exact Design of Band-stop Microwave Filters," *IEEE Trans., Microwave Theory and Techniques,* MTT-12, No. 1, January 1964, pp. 6–15.

[20] Malherbe, J.A.G., *Microwave Transmission Line Filters,* Norwood, MA: Artech House, 1979.

[21] Chen, W.K., *Theory and Design of Broadband Matching Networks,* New York: Pergamon Press, 1976.

[22] Przedpełski, A.B., "Bandwidth of Transmission Line Matching Circuits," *Microwave Journal,* Vol. 21, No. 4, April 1978, pp. 71–73 and 76.

[23] Gupta, K.C. and R. Chadha, "Design Real-World Stripline Circuits," *Microwaves,* Vol. 17, No. 12, December 1978, pp. 70–80.

[24] Rosłoniec, S., "Design of Transmission Line Matching Circuits," *RF Design,* Vol. 13, No. 2, February 1990, pp. 52–56.

Chapter 6
SINGLE AND COUPLED TRANSMISSION LINES

6.1 INTRODUCTION

In this chapter, we discuss the fundamental transverse electromagnetic TEM (or quasi-TEM) single and coupled transmission lines, which are widely used in radio frequency and microwave equipment [1,2]. The algorithms for designing the single eccentric coaxial line, slab line, triplate stripline, microstrip line and inverted microstrip line are described in Sections 6.2, 6.3, 6.4, 6.5 and 6.6, respectively. The algorithms for the design of coupled slab lines, coupled triplate striplines and coupled microstrip lines are covered in sections 6.7, 6.8 and 6.9. We note that these algorithms are implemented in the accompanying computer programs, which we present with check calculations.

In all the design algorithms presented, particular emphasis is put on their accuracy and reliability. Hence, in most computational programs, the short procedures are included for protection from different kinds of errors. The essence of these check and protection procedures is explained in Chapter 3. Moreover, it was our intention to make the computational programs as fast as possible.

As we discussed earlier, the computer programs given in this chapter form the Library L3 of the MICRO system. These programs can be run either directly from the procedure MENU (see Figure 3.1) or indirectly (by means of procedure CONTROL) as subroutines.

6.2 THE ECCENTRIC COAXIAL LINE

Let us consider the coaxial line, the inner conductor of which is laterally displaced from its normal position to the axis location, as shown in the transverse section view in Figure 6.1. The characteristic impedance of this TEM transmission line can be evaluated by a simple field analysis [1] and the resulting expression is

$$Z_0(x) = 59.952 \sqrt{\frac{\mu_r}{\epsilon_r}} \ln(x + \sqrt{x^2 - 1}), \Omega \tag{6.1}$$

where:

ϵ_r — the relative permittivity of the dielectric substrate,
μ_r — the relative permeability of the dielectric substrate,
$x = [b + (a^2 - 4c^2)/b]/(2a)$,

and the structural parameters a, b, and c are defined in Figure 6.1.

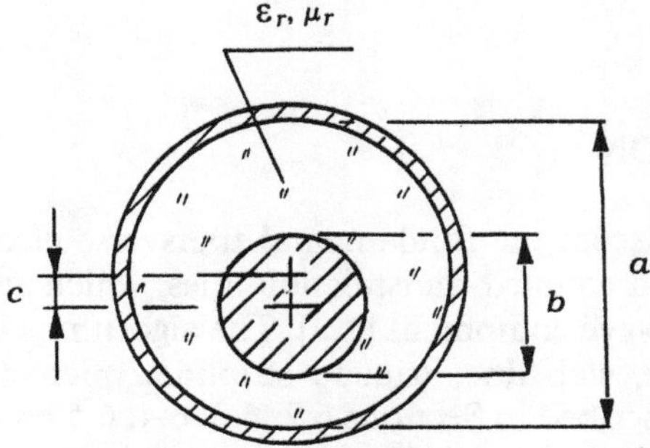

Figure 6.1 Cross section of the eccentric coaxial line.

We can easily see that when $c = 0$, the line under discussion becomes the normal coaxial line for which

$$Z_0(a/b) = 59.952 \sqrt{\frac{\mu_r}{\epsilon_r}} \ln(a/b), \Omega \tag{6.2}$$

When making practical use of the data given above, we must bear in mind the restrictions concerning the higher-order (waveguide) modes. According to Gunston [2], the pure TEM mode can only be assured if the coaxial line dimensions and the operating frequency are chosen such that the propagation wavelength λ obeys the inequality

$$\lambda > \pi(a + b)/2$$

The design of the eccentric coaxial line for given values of Z_0, a, c, ϵ_r and μ_r consists of evaluating the suitable diameter b of the inner conductor. From the standpoint of mathematics, this is equivalent to solving the equation

$$Z_0(b) - Z_0 = 0$$

which in the ECL program presented below is solved by the golden section method. See Section 2.2.3 or [3,4].

Example of Calculations

Some results of calculations obtained by using (6.1) and (6.2) are given in Table 6.1.

Table 6.1
Geometrical Dimensions of the Eccentric Coaxial Line as a Function of Its Characteristic Impedance

	$\epsilon_r = 1, \mu_r = 1$		
Z_0, Ω	a, m	b, m	c, m
50.000	$7 \cdot 10^{-3}$	$3.040 \cdot 10^{-3}$	0
50.000	$7 \cdot 10^{-3}$	$3.038 \cdot 10^{-3}$	$1 \cdot 10^{-4}$
50.000	$7 \cdot 10^{-3}$	$3.026 \cdot 10^{-3}$	$2 \cdot 10^{-4}$
49.939	$7 \cdot 10^{-3}$	$3.040 \cdot 10^{-3}$	$1 \cdot 10^{-4}$
49.758	$7 \cdot 10^{-3}$	$3.040 \cdot 10^{-3}$	$2 \cdot 10^{-4}$
49.453	$7 \cdot 10^{-3}$	$3.040 \cdot 10^{-3}$	$3 \cdot 10^{-4}$

Program ECL

Program ECL allows us to calculate the diameter b of the inner conductor of the eccentric coaxial line and normal coaxial line ($c = 0$) if the values of the following input data are known:

Z_0 — the characteristic impedance in ohms,
ϵ_r — the relative permittivity of the dielectric substrate,
μ_r — the relative permeability of the dielectric substrate,
a — the diameter of the outer conductor in m,
c — the displacement of the inner conductor in m.

Listing of Program ECL

```
7700 REM ECL
7702 CLS
7704 PRINT "ECCENTRIC COAXIAL LINE"
7706 PRINT
7708 PRINT "Data:"
7710 PRINT
7712 INPUT "Zo[ohm]=",ZO
```

```
7714 INPUT "er=",ER
7716 INPUT "ur=",UR
7718 INPUT "a[m]=",A
7720 INPUT "c[m]=",C
7722 IF ZO <= 0 THEN  PRINT "Error:Zo <= 0": GOTO 7710
7724 IF ER<1 THEN  PRINT "Error:er<1": GOTO 7710
7726 IF UR <= 0 THEN  PRINT "Error:ur <= 0": GOTO 7710
7728 PRINT
7730 LET ZC=ZO
7732 GOSUB 7742
7734 PRINT "Results, see Fig. 6.1"
7736 PRINT
7738 PRINT "b[m]=";B
7740 GOTO 7792
7742 REM Procedure ECL
7744 IF C<0 OR C >= A/2 THEN  PRINT "Wrong value of offset c (
0 <= c<";A/2;" )": PRINT : GOTO 7720
7746 IF C=0 THEN  GOTO 7780
7748 LET B1=.01*A
7750 LET B2=1.99*(A/2-C)
7752 LET B=B1
7754 GOSUB 7786
7756 LET Z1=Z
7758 LET B=B2
7760 GOSUB 7786
7762 LET Z2=Z
7764 IF Z1*Z2>0 THEN  PRINT "The line cannot be realized !": ST
OP : GOTO 7712
7766 LET B=(B1+B2)/2
7768 GOSUB 7786
7770 IF  ABS (Z) <= ZC/500 THEN  GOTO 7734
7772 IF Z1*Z>0 THEN  LET B1=B: LET Z1=Z: GOTO 7766
7774 LET B2=B
7776 LET Z2=Z
7778 GOTO 7766
7780 LET X=ZC* SQR (ER/UR)/59.952
7782 LET B=A/( EXP(X))
7784 RETURN
7786 LET X=(B/A+(A*A-4*C*C)/(A*B))/2
7788 LET Z=59.952* SQR (UR/ER)* LOG (X+ SQR (X*X-1))-ZC
7790 RETURN
7792 REM Chain subroutine
7794 PRINT
7796 INPUT"Enter 'CONT':",N$
7798 IF N$="CONT" THEN CLS:CHAIN "MENU",10
7800 GOTO 7794
```

6.3 THE SLAB LINE

The transverse section of an unscreened slab line is shown in Figure 6.2. This kind of TEM transmission line is widely used in the microwave filter technology because their use offers several manufacturing advantages and excellent electrical filter properties [17,19]. The air slab line is also used in measurement techniques as an easily-accessible structure for testing standing waves. Therefore, its analysis attracted considerable attention, and some of the important results in this area have been published [2,5 and 7]. Gunston [2] has concluded that Wheeler's analysis formula [5] is the most accurate. The advantage of this formula lies also in its simple mathematical form, namely:

$$Z_0(d/b) = 59.952 \sqrt{\frac{\mu_r}{\epsilon_r}} \left(\ln \frac{\sqrt{X} + \sqrt{Y}}{\sqrt{X - Y}} - \frac{R^4}{30} + 0.014R^8 \right), \Omega \tag{6.3}$$

where:

$R = \pi d/(4b)$,
$X = 1 + 2 \sinh^2(R)$,
$Y = 1 - 2 \sin^2(R)$,
ϵ_r = the relative permittivity of the dielectric substrate,
μ_r = the relative permeability of the dielectric substrate.

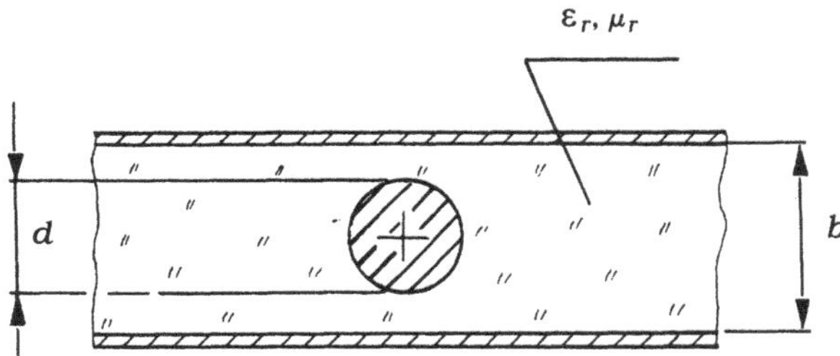

Figure 6.2 Cross section of the single slab line.

The synthesis of the slab line on the basis of (6.3) consists of evaluating such a ratio d/b (see Figure 6.2), for which the following equation is satisfied:

$$V(d/b) = |Z_0(d/b) - Z_0| = 0$$

where Z_0 is the characteristic impedance of the line being designed. The function $V(d/b)$ assumes its minimum value (zero) at point (d/b). This point solution can be effectively evaluated by means of different minimization methods, for instance, the bisection, Fibonacci search, or search by the golden section [4,6].

Interesting results of analysis of the slab line with an anisotropic dielectric substrate have been published [7]. Let us assume that the characteristic impedance

Z_0 and tensor $\bar{\bar{\epsilon}}_r$ of the dielectric substrate are given. In this case, the design of the line consists of solving the following nonlinear equation:

$$Z_0(b/d, \bar{\bar{\epsilon}}_r) - Z_0 = 0 \tag{6.4}$$

where

$$Z_0(b/d, \bar{\bar{\epsilon}}_r) = \frac{30\pi}{\sqrt[4]{\epsilon_x \epsilon_y}} \frac{K'(k)}{K(k)}, \ \Omega, \tag{6.5}$$

$$\frac{b}{d} = \frac{\pi}{\sqrt{\epsilon_x/\epsilon_y} \ln[(1 + k)/(1 - k)]} + \frac{\pi}{2 \sin^{-1}(k)},$$

$$\bar{\bar{\epsilon}}_r = \begin{bmatrix} \epsilon_x, & 0 \\ 0, & \epsilon_y \end{bmatrix}$$

$K(k)$ = the complete elliptic integral of the first kind,
$K'(k)$ = the complementary to $K(k)$, complete elliptic integral of the first kind (see Appendix 1),
k = the modulus, $0 \leqslant k \leqslant 1$.

In order to solve (6.4) with respect to ratio b/d, we must evaluate in turn: the ratio of the elliptic integrals $K'(k)/K(k)$, the modular constant $q = \exp[-\pi K'(k)/K(k)]$ and the modulus k. If the value of the modular constant q is known, then the modulus k can be calculated by using the following series [8]:

$$k = 4\sqrt{q}\left[\frac{(1 + q^2)(1 + q^4)(1 + q^6)\ldots}{(1 + q)(1 + q^3)(1 + q^5)\ldots}\right]^4 \tag{6.6}$$

Next, the value of b/d is calculated (see (6.5)).

The above formulas are used in the computer program SL. If $\epsilon_x = \epsilon_y = \epsilon_r$, the calculations are made on the basis of (6.3). In this case, the design problems are solved by two numerical methods, namely the linear search and search by the golden section. We note that for $\epsilon_x = \epsilon_y = \epsilon_r$, (6.5) is not applied in program SL because its calculations are not always accurate, especially for the smallest values of Z_0 and ϵ_r.

Check Calculations

Some numerical results obtained by using (6.3) and (6.5) are presented in Table 6.2.

Table 6.2
Geometrical Dimensions of the Single Slab Line as a Function of Its Characteristic Impedance

			d/b	
Z_0, Ω	ϵ_x	ϵ_y	*(6.3)*	*(6.5)*
30	1.00	1.00	0.7487	0.7199
30	3.78	3.78	0.4790	0.4760
50	1.00	1.00	0.5486	0.5423
50	3.78	3.78	0.2515	0.2517
75	1.00	1.00	0.3639	0.3633
75	3.78	3.78	0.1118	0.1119
50	11.60	9.40	—	0.0906
50	5.12	3.40	—	0.2544

Program SL

The input data for the computer program SL are:

Z_0 = the characteristic impedance of the slab line being designed (in ohms),
ϵ_x = the component of the permittivity tensor $\bar{\bar{\epsilon}}_r$,
ϵ_y — the component of the permittivity tensor $\bar{\bar{\epsilon}}_r$,
μ_r — the relative permeability of the dielectric substrate.

From the calculations, we find the value of d/b (see Figure 6.2).

Listing of Program SL

```
7900 REM SL
7902 CLS
7904 PRINT "SLAB LINE"
7906 PRINT
7908 PRINT "Data:"
7910 PRINT
7912 INPUT "Zo[ohm]=",ZO
7914 INPUT "ex=",EX
7916 INPUT "ey=",EY
7918 INPUT "ur=",UR
7920 PRINT
7922 PRINT "Please wait !"
7924 PRINT
7926 LET ZC=ZO
7928 LET PI=3.141593
7930 IF ZC <= 0 THEN  PRINT "Error:Zo <= 0": GOTO 7910
```

```
7932 IF EX<1 THEN  PRINT "Error:ex<1": GOTO 7910
7934 IF EY<1 THEN  PRINT "Error:ey<1": GOTO 7910
7936 IF UR <= 0 THEN  PRINT "Error:ur <= 0": GOTO 7910
7938 IF EX=EY THEN  LET ER=EX: GOTO 7972
7940 IF UR=1 THEN  GOTO 7944
7942 PRINT "The formula (6.5) holds only for ur=1": GOTO 7906
7944 LET C=ZC* SQR ( SQR (EX*EY))/30
7946 LET Q= EXP (-C)
7948 LET IL=1
7950 FOR I=1 TO 10
7952 LET IL=IL*(1+Q^(2*I))/(1+Q^(2*I-1))
7954 NEXT I
7956 LET K=4* SQR (Q)*IL*IL*IL*IL
7958 LET G= SQR (EX/EY)* LOG ((1+K)/(1-K))
7960 LET H=2* ATN (K/SQR(1-K*K))
7962 LET S= PI *(1/G+1/H)
7964 PRINT "Results, see Fig. 6.2"
7966 PRINT
7968 PRINT "d/b=";1/S
7970 GOTO 8030
7972 REM Procedure SL
7974 GOSUB 7980
7976 LET S=1/V
7978 GOTO 7964
7980 REM Subroutine 7980
7982 LET V1=.01
7984 LET V=V1
7986 GOSUB 8018
7988 LET F1=F
7990 LET I=1
7992 LET H=.2
7994 LET V=V1+I*H
7996 GOSUB 8018
7998 IF F1*F <= 0 THEN  GOTO 8004
8000 LET I=I+1
8002 GOTO 7994
8004 LET V2=V
8006 LET V=(V1+V2)/2
8008 GOSUB 8018
8010 IF  ABS (F) <= ZC/1000 THEN  RETURN
8012 IF F1*F>0 THEN  LET V1=V: GOTO 8006
8014 LET V2=V
8016 GOTO 8006
8018 REM Subroutine 8018
8020 LET R= PI *V/4
8022 LET X=1+( EXP (R)- EXP (-R))^2/2
8024 LET Y=1-2*( SIN (R))^2
8026 LET F=59.952* SQR (UR/ER)*( LOG (( SQR (X)+ SQR (Y))/ SQR
(X-Y))-R^4/30+.01*R^8)-ZC
8028 RETURN
8030 REM Chain subroutine
8032 PRINT
8034 INPUT"Enter 'CONT':",N$
8036 IF N$="CONT" THEN CLS:CHAIN "MENU",10
8038 GOTO 8032
```

6.4 THE TRIPLATE STRIPLINE

The triplate stripline, the cross section of which is shown in Figure 6.3, is one of the TEM transmission lines most commonly used at very high frequencies. Usually stripline circuits are constructed from copper-clad printed-circuit boards, where the center conductor thickness t is very small in comparison with other transverse dimensions of the line. Analysis of the line being considered is virtually simplified when the thickness t is negligible (i.e., when t is equal to zero). In this case, the characteristic impedance Z_0 is given by the exact closed formula [9]:

$$Z_0(W/b) = 29.976\pi \sqrt{\frac{\mu_r}{\epsilon_r}} \cdot \frac{K(k)}{K'(k)}, \ \Omega \tag{6.7}$$

where:

$k = 1/\cosh[\pi W/(2b)]$

ϵ_r = the relative permittivity of the dielectric substrate,

μ_r = the relative permeability of the dielectric substrate,

$K(k)$ = the complete elliptic integral of the first kind,

$K'(k)$ = the complementary to $K(k)$ complete elliptic integral of the first kind (see Appendix 1).

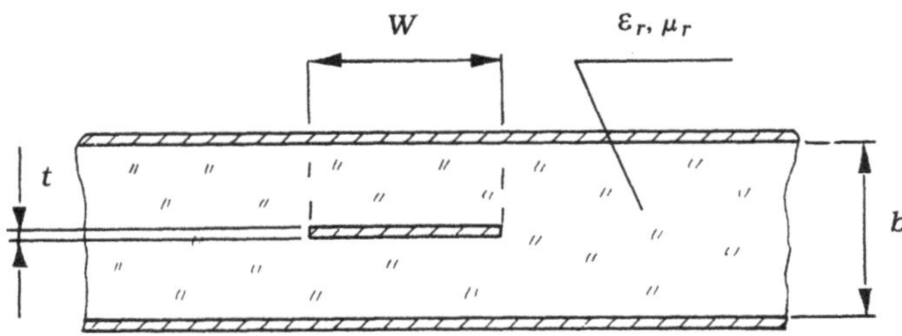

Figure 6.3 Cross section of the triplate stripline.

For evaluating the geometrical dimensions of the line being designed, when its electrical parameters such as Z_0, ϵ_r and μ_r are known, we must calculate the modulus k of the elliptic integrals $K(k)$ and $K'(k)$. We can see that the ratio of these integrals is found from (6.7). Hence, the modulus k can be calculated in a manner similar to that employed in Section 6.3 (see (6.6)), or on the basis of the following series [10]:

$$k = \sqrt{q}\,(N/D)^2 \tag{6.8}$$

where:

$$q = \exp[-\pi K'(k)/K(k)]$$

$$N = \sum_{i=1}^{\infty} q^{i(i-1)}$$

$$D = 0.5 + \sum_{i=1}^{\infty} q^{i \cdot i}$$

Here we note that the series in (6.8) is rapidly convergent, and accuracy of calculations is assured at $i = 6$ and $0.0006 < q < 0.27$. Having thus found a value of the modulus k, we can calculate the width of the strip from the formula:

$$W = \frac{2b}{\pi} \ln(1/k + \sqrt{1/k^2 - 1})$$

In the literature, [6] researchers have concluded that for $2 \leqslant \epsilon_r \leqslant 10$, $20 \leqslant Z_0 \leqslant 100\ \Omega$ and $t \leqslant 0.001 \cdot b$, (6.7) ensures an accuracy not worse than 0.5 percent for impedance Z_0. In other cases, we must take into account the finite thickness t of the central conductor. For this purpose, we can use the closed form expressions reported by Wheeler [11]. Then:

$$Z_0(W/b, t) = \frac{30}{\sqrt{\epsilon_r}} \ln[1 + A(2A + \sqrt{4A^2 + 6.27})]\ \Omega \tag{6.9}$$

where:

$$A = 4(b - t)/[\pi(W + \Delta W)],$$

$$\Delta W = \frac{t}{\pi}\left\{1 - \frac{1}{2}\ln\left[\left(\frac{t}{2b - t}\right)^2 + \left(\frac{0.0796t}{W + 1.1t}\right)^m\right]\right\},$$

$$m = 6(b - t)/(3b - t).$$

For $A > 0.128$, (6.9) is said to yield data which are accurate within 0.5 percent.

The expressions for the synthesis of the line, derived from (6.9) are:

$$W = W_0 - \Delta W_0 \tag{6.10}$$

where

$$W_0 = \frac{8(b-t)}{\pi}\frac{\sqrt{B+0.568}}{B-1}$$

$$\Delta W_0 = \frac{t}{\pi}\left\{1 - \frac{1}{2}\ln\left[\left(\frac{t}{2b-t}\right)^2 + \left(\frac{0.0796t}{W_0 - 0.26t}\right)^m\right]\right\}$$

$$B = \exp(Z_0\sqrt{\epsilon_r}/30)$$

$$m = 6(b-t)/(3b-t).$$

The design formulas given above are used in the computer program TL, which is presented below. If $t < 0.001 \cdot b$, the calculations are made on the basis of (6.7). The lines with $t \geqslant 0.001 \cdot b$ are designed by using (6.10).

Check Calculations

Some computational results obtained from (6.7), (6.9) and (6.10) are given in Tables 6.3 and 6.4.

Table 6.3
Characteristic Impedance of the Triplate Stripline as a Function of its Geometrical Dimensions (The results presented have been obtained by using (6.7) and (6.9))

		$b = 2\ 10^{-3}$ *m*, $t = 0$ *m*	
W, m	ϵ_r	Z_0, Ω (6.7)	Z_0, Ω (6.9)
$3.049 \cdot 10^{-3}$	2.55	30.00	29.89
$1.088 \cdot 10^{-3}$	10.20	30.00	29.99
$1.478 \cdot 10^{-3}$	2.55	50.00	49.95
$0.375 \cdot 10^{-3}$	10.20	50.00	50.03
$0.708 \cdot 10^{-3}$	2.55	75.00	75.03
$0.094 \cdot 10^{-3}$	10.20	75.00	75.06

Table 6.4
Geometrical Dimensions of the Finite-Thickness Triplate Stripline as a Function of its Characteristic Impedance (The results presented have been obtained by using (6.10))

			$b = 2\ 10^{-3}$ *m*
Z_0, Ω	ϵ_r	*t, m*	*W, m*
30.00	2.55	$3 \cdot 10^{-5}$	$2.933 \cdot 10^{-3}$
30.00	10.20	$1 \cdot 10^{-5}$	$1.060 \cdot 10^{-3}$
50.00	2.55	$3 \cdot 10^{-5}$	$1.397 \cdot 10^{-3}$
50.00	10.20	$1 \cdot 10^{-5}$	$0.335 \cdot 10^{-3}$
75.00	2.55	$3 \cdot 10^{-5}$	$0.643 \cdot 10^{-3}$
75.00	10.20	$1 \cdot 10^{-5}$	$0.075 \cdot 10^{-3}$

Program TL

The program TL is used for designing the triplate striplines. The input data for this program are:

Z_0 = the characteristic impedance of the line being designed in ohms,
ϵ_r = the relative permittivity of the dielectric substrate,
μ_r = the relative permeability of the dielectric substrate,
b = the thickness of the dielectric substrate in m,
t = the thickness of the strip in m.

From the calculations, we obtain the width W of the strip, as shown in Figure 6.3.

Listing of Program TL

```
8200 REM TL
8202 CLS:DEFDBL A-Z:DEFINT I
8204 PRINT "TRIPLATE STRIPLINE"
8206 PRINT
8208 PRINT "Data:"
8210 PRINT
8212 INPUT "Zo[ohm]=",ZO
8214 INPUT "er=",ER
8216 INPUT "ur=",UR
8218 INPUT "b[m]=",B
8220 INPUT "t[m]=",T
8222 PRINT
8224 LET PI=3.141592653#
8226 IF ZO <= 0 THEN  PRINT "Error:Zo <= 0": GOTO 8210
8228 IF ER<1 THEN  PRINT "Error: er<1": GOTO 8210
8230 IF UR <= 0 THEN  PRINT "Error: ur <= 0": GOTO 8210
8232 IF B <= 0 THEN  PRINT "Error:b <= 0": GOTO 8210
8234 IF T>B/10 THEN  PRINT "t>b/10": GOTO 8210
8236 PRINT "Please wait !"
8238 PRINT
8240 LET ZC=ZO
8242 GOSUB 8252
8244 PRINT "Results, see Fig. 6.3"
8246 PRINT
8248 PRINT USING "W[m]=###.######";W
8250 GOTO 8292
8252 REM Procedure TL
8254 IF T >= B/1000 THEN  GOTO 8276
8256 LET Q= EXP (-29.976* PI * PI /ZC* SQR (UR/ER))
8258 LET L=0
8260 LET M=.5
8262 FOR I=1 TO 6
8264 LET L=L+Q^(I*(I-1))
```

```
8266 LET M=M+Q^(I*I)
8268 NEXT I
8270 LET K= SQR (Q)*(L/M)*(L/M)
8272 LET W=B*2/ PI * LOG (1/K+ SQR (1/(K*K)-1))
8274 RETURN
8276 IF UR=1 THEN  GOTO 8280
8278 PRINT "The formula (6.10) holds only for ur=1": GOTO 8206
8280 LET BL= EXP (ZC* SQR (ER)/30)
8282 LET M=6*(B-T)/(3*B-T)
8284 LET WO=8*(B-T)* SQR (BL+.568)/( PI *(BL-1))
8286 LET WC=T*(1-.5* LOG (T/(2*B-T)*T/(2*B-T)+(.0796*T/(WO-.26*
T))^M))/ PI
8288 LET W=WO-WC
8290 RETURN
8292 REM Chain subroutine
8294 PRINT
8296 INPUT"Enter 'CONT':",N$
8298 IF N$="CONT" THEN CLS:CHAIN "MENU",10
8300 GOTO 8294
```

6.5 THE MICROSTRIP LINE

Let us consider the microstrip transmission line, a transverse section of which is shown in Figure 6.4. This transmission line is a nonhomogeneous structure because the lines of the electric field, between the strip and the ground plane, are not contained entirely in the dielectric substrate. Therefore, the wave-mode propagating along the microstrip is not purely TEM but quasi-TEM. The effective dielectric constant ϵ_{eff} of the line takes into account the field external to the dielectric substrate, and is lower than the relative permittivity ϵ_r.

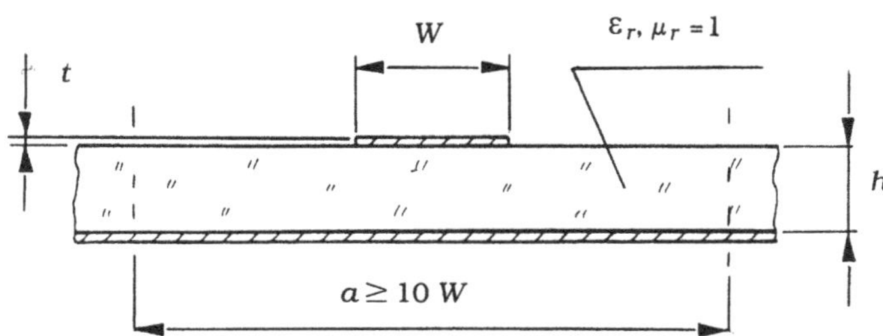

Figure 6.4 Cross section of the microstrip line.

Atwater [12] has found that from the point of view of accuracy, the closed form expressions for characteristic impedance Z_0 and effective dielectric constant ϵ_{eff} given in [13] and [14] are most useful for computer-aided design. If the values of Z_0, h, ϵ_r, t and frequency f (see Figure 6.4) are known, then designing the microstrip line is equivalent to solving the following equation:

$$Z_0(u, f) - Z_0 = 0 \tag{6.11}$$

where:

$$u = W/h$$

$$Z_0(u, f) = \frac{60}{\sqrt{\epsilon_{\text{eff}}(f)}} \ln\left[\frac{f(u)}{u} + \sqrt{1 + \left(\frac{2}{u}\right)^2}\right], \Omega \tag{6.12}$$

$$\epsilon_{\text{eff}}(u, f) = \frac{\epsilon_{\text{eff}}(u, 0) + \epsilon_r p(u, f)}{1 + p(u, f)}$$

$$\epsilon_{\text{eff}}(u, 0) = \frac{\epsilon_r + 1}{2} + \frac{\epsilon_r - 1}{2}\left(1 + \frac{10}{u}\right)^{-a(u)b(\epsilon_r)}$$

$$a(u) = 1 + \frac{1}{49}\ln\left[\frac{u^4 + (u/52)^2}{u^4 + 0.432}\right] + \frac{1}{18.7}\ln\left[1 + \left(\frac{u}{18.1}\right)^3\right]$$

$$b(\epsilon_r) = 0.564[(\epsilon_r - 0.9)/(\epsilon_r + 3)]^{0.053}$$

$$f(u) = 6 + (2\pi - 6)\exp[-(30.666/u)^{0.7528}]$$

$$p(u, f) = n(u, f)c(\epsilon_r)[1.844f_n + k(u, f)\, d(\epsilon_r)f_n]^{1.5763}$$

$$c(\epsilon_r) = 0.33622[1 - \exp(-0.03442\epsilon_r)]$$

$$f_n = 10^{-7} \cdot f \cdot h$$

$$d(\epsilon_r) = 3.751 - 2.75 \cdot \exp[-(\epsilon_r/15.916)^8]$$

$$k(u, f) = 0.363g(f)\exp(-4.6u)$$

$$g(f) = 1 - \exp[-(f_n/3.87)^{4.97}]$$

$$n(u, f) = 0.27488 + 0.6315u + m(f)u - 0.065683 \cdot \exp(-8.7513u)$$

$$m(f) = 0.525/(1 + 0.157f_n)^{20}$$

In computer program ML, presented below, (6.11) is solved by using two numerical methods (i.e., the linear search (step by step) and the search by golden section [4,6]). In the first stage of the process, the interval $(u_i, u_i + \Delta u)$ (see Figure 6.5a) including the solution u_0, is determined. At points u_i and $u_i + \Delta u$, function $V(u, f) = Z_0(u, f) - Z_0$ takes values which have different signs. Then, by the golden section method, the position of the minimum of the function $|V(u, f)|$ is found (see Figure 6.5(b)). The search of the position of the solution u_0 terminates if

$$|V(u, f)| < Z_0/10{,}000$$

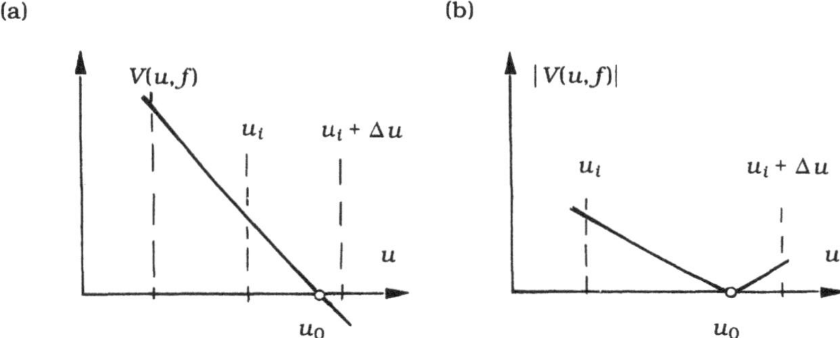

Figure 6.5 Illustration of the process of solving (6.11).

The solution obtained in this way refers to a microstrip line with the strip thickness $t = 0$. The nonzero thickness of the strip and the dielectric discontinuity air-dielectric substrate may be accounted for by subtracting from the solution u_0 the correction Δu [13]:

$$\Delta u = \frac{t}{2\pi h} \ln \left(1 + \frac{4eh}{t \coth^2 \sqrt{6.517\, u}} \right) \left(1 + \frac{1}{\cosh \sqrt{\epsilon_r - 1}} \right)$$

Check Calculations

As we said earlier, the design formulas presented above are used in computer program ML. Some check results obtained by this program are given in Table 6.5.

Table 6.5
Some Results of Design of the Microstrip Line

			$h = 10^{-3}\ m,\ t = 3\ 10^{-5}\ m$	
Z_0, Ω	ϵ_r	f, Hz	W, m	ϵ_{eff}
30	2.55	10^6	$5.678 \cdot 10^{-3}$	2.228
30	2.55	10^{10}	$5.558 \cdot 10^{-3}$	2.299
30	9.60	10^6	$2.395 \cdot 10^{-3}$	7.042
30	9.60	10^{10}	$2.201 \cdot 10^{-3}$	7.772
50	2.55	10^6	$2.762 \cdot 10^{-3}$	2.122
50	2.55	10^{10}	$2.705 \cdot 10^{-3}$	2.177
50	9.60	10^6	$0.961 \cdot 10^{-3}$	6.448
50	9.60	10^{10}	$0.877 \cdot 10^{-3}$	6.972
75	2.55	10^6	$1.379 \cdot 10^{-3}$	2.033
75	2.55	10^{10}	$1.352 \cdot 10^{-3}$	2.073
75	9.60	10^6	$0.339 \cdot 10^{-3}$	6.067
75	9.60	10^{10}	$0.307 \cdot 10^{-3}$	6.431

Program ML

The program ML allows us to calculate the geometrical dimensions of the micro-strip line if the following input data are known:

Z_0 = the characteristic impedance in ohms,
ϵ_r = the relative permittivity of the dielectric substrate,
$\mu_r = 1$
h = The thickness of the dielectric substrate in m,
t = the thickness of the strip in m,
f = the signal frequency in Hz.

From the calculations, we find the width W of the strip and the effective dielectric constant ϵ_{eff}.

Listing of Program ML

```
8400 REM ML
8402 CLS:DEFDBL A-Z
8404 PRINT "MICROSTRIP LINE"
8406 PRINT
8408 PRINT "Data:"
8410 PRINT
8412 INPUT "Zo[ohm]=",ZO
8414 INPUT "er=",ER
8416 INPUT "h[m]=",H
8418 INPUT "t[m]=",T
8420 INPUT "f[Hz]=",F
8422 LET ZC=ZO
8424 LET PI=3.141592653#
8426 IF ZC <= 0 THEN  PRINT "Error:Zo <= 0": GOTO 8410
8428 IF ER<1 THEN  PRINT "Error: er<1": GOTO 8410
8430 IF H <= 0 THEN  PRINT "Error: h <= 0": GOTO 8410
8432 IF T<0 THEN  PRINT "Error: t<0": GOTO 8410
8434 IF F <= 0 THEN  PRINT "Error: f <= 0": GOTO 8410
8436 PRINT
8438 PRINT "Please wait !"
8440 PRINT
8442 GOSUB 8454
8444 PRINT "Results, see Fig. 6.4"
8446 PRINT
8448 PRINT USING "ef=##.######";EF
8450 PRINT USING "W[m]=##.######";(U-DTU)*H
8452 GOTO 8580
8454 REM Procedure ML
8456 LET B=.564*((ER-.9)/(ER+3))^.053
8458 LET C=.33622*(1- EXP (-.03442*ER))
8460 LET D=3.751-2.751* EXP (-(ER/15.916)^8)
8462 LET Q=.0000001*F*H
```

```
8464 LET G=1- EXP (-(Q/3.87)^4.97)
8466 LET M=.525/(1+.157*Q)^20
8468 LET U=.1
8470 LET DU=.5
8472 GOSUB 8558
8474 LET V1=V
8476 LET U=U+DU
8478 GOSUB 8558
8480 LET V2=V
8482 IF V1*V2 <= 0 THEN  GOTO 8488
8484 LET V1=V2
8486 GOTO 8476
8488 LET U2=U
8490 LET U1=U-DU
8492 LET U=U1+(U2-U1)*.381966
8494 GOSUB 8558
8496 LET U3=U
8498 LET V3= ABS (V)
8500 LET U=U1+(U2-U1)/1.618034
8502 GOSUB 8558
8504 LET U4=U
8506 LET V4= ABS (V)
8508 IF (V3+V4) <= ZC/20000 THEN  LET U=(U1+U2)/2: GOTO 8544
8510 IF V3>V4 THEN  GOTO 8528
8512 LET U2=U4
8514 LET U4=U3
8516 LET V4=V3
8518 LET U=U1+(U2-U1)*.381966
8520 GOSUB 8558
8522 LET U3=U
8524 LET V3= ABS (V)
8526 GOTO 8508
8528 LET U1=U3
8530 LET U3=U4
8532 LET V3=V4
8534 LET U=U1+(U2-U1)/1.618034
8536 GOSUB 8558
8538 LET U4=U
8540 LET V4= ABS (V)
8542 GOTO 8508
8544 IF T=0 THEN  LET DTU=0: GOTO 8444
8546 LET X= SQR (U*6.517)
8548 LET CTH=( EXP (2*X)+1)/( EXP (2*X)-1)
8550 LET X= SQR (ER-1)
8552 LET CH=( EXP (X)+ EXP (-X))/2
8554 LET DTU= LOG (1+4*H* EXP (1)/(T*CTH*CTH))*(1+1/CH)*T/(2*H*
 PI )
8556 RETURN
8558 REM Subroutine 8558
8560 LET A=1+1/49* LOG ((U^4+(U/52)^2)/(U^4+.432))+1/18.7* LOG
(1+(U/18.1)^3)
8562 LET F=6+(2* PI -6)* EXP (-(30.666/U)^.7528)
8564 LET K=.363*G* EXP (-4.6*U)
8566 LET N=.27488+.6315*U+M*U-.065683* EXP (-8.7513*U)
8568 LET P=N*C*(1.844*Q+K*D*Q)^1.5763
```

```
8570 LET EO=(ER+1)/2+(ER-1)/2*(1+10/U)^(-A*B)
8572 LET EF=(EO+ER*P)/(1+P)
8574 LET Z=60/ SQR (EF)* LOG (F/U+ SQR (1+(2/U)^2))
8576 LET V=Z-ZC
8578 RETURN
8580 REM Chain subroutine
8582 PRINT
8584 INPUT"Enter 'CONT':",N$
8586 IF N$="CONT" THEN CLS:CHAIN "MENU",10
8588 GOTO 8582
```

6.6 THE INVERTED MICROSTRIP LINE

The inverted microstrip line is one of the principal transmission media used in the upper microwave and lower millimeter-wave bands. The most interesting aspect of this line is that the presence of an air gap between the ground plane and the dielectric substrate reduces the effects of dispersion on the propagation constant. For this reason, the quasistatic results of the analysis remain useful even at very high frequencies [15,16].

The cross-section of the line being considered is shown in Figure 6.6, with parameters W, d, h, ϵ_r and t (strip thickness) defined therein. According to [15], the characteristic impedance Z_0 and effective dielectric constant ϵ_{eff} can be computed by using the following closed form expressions:

$$Z_0(u) = \frac{60}{\sqrt{\epsilon_{\mathrm{eff}}}} \ln \left[\frac{f(u)}{u} + \sqrt{1 + \left(\frac{2}{u}\right)^2} \right] \ \Omega$$

$$\sqrt{\epsilon_{\mathrm{eff}}} = 1 + \frac{d}{h}[c_1 - c_2 \ln(u)](\sqrt{\epsilon_r} - 1) \tag{6.13}$$

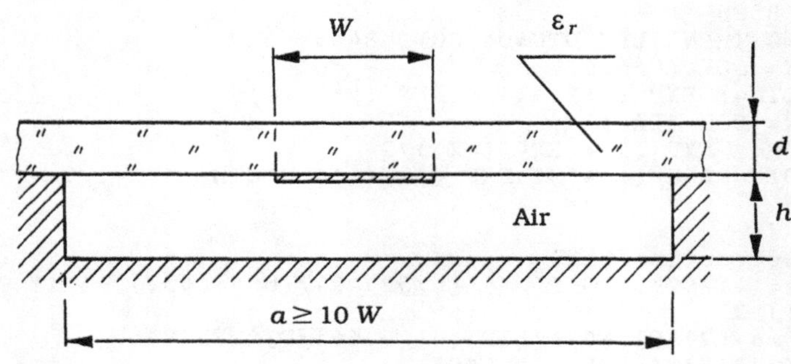

Figure 6.6 Cross section of the inverted microstrip line.

where:

$$u = W/h$$
$$f(u) = 6 + (2\pi - 6)\exp[-(30.666/u)^{0.7528}],$$
$$c_1 = [0.5173 - 0.1515\ln(d/h)]^2,$$
$$c_2 = [0.3092 - 0.1047\ln(d/h)]^2$$

Let us assume that the values of Z_0, ϵ_r, d and h are known. Then the design of the inverted microstrip line consists of solving the equation:

$$Z_0(u) - Z_0 = 0 \tag{6.14}$$

Here, we note that (6.14) is nonlinear with respect to $u = W/h$, different numerical methods can be applied to solve it. In computer program IML, given below, this equation is solved in a manner similar to that employed in program ML (see Section 6.5) (i.e., by the linear search and search by the golden section).

In [15], Pramanick and Bhartia have shown that the formulas presented above ensure good agreement of calculations with the corresponding exact results published in the literature. The relative deviations of the characteristic impedance Z_0 and effective dielectric constant are smaller than 1 percent for $0.2h \leqslant d \leqslant h$, $h \leqslant W \leqslant 8h$, $0 < t \leqslant 0.02h$ and $\epsilon_r \leqslant 6$. If these constraints are not satisfied, we have to apply the more general, but also more complicated design formulas which have been recently published [16].

Check Calculations

Some numerical results obtained by running computer program IML are given in Table 6.6.

Table 6.6
Some Results of Design of the Inverted Microstrip Line

	$d = 5\ 10^{-4}$ *m*, $h = 10^{-3}$ *m*, $t = 10^{-5}$ *m*, $\mu_r = 1$			
Z_0, Ω	ϵ_r	W/h	ϵ_{eff}	*Comments*
30	2.55	9.399	1.036	$W/h > 8$
30	3.78	9.206	1.061	$W/h > 8$
50	2.55	4.599	1.100	
50	3.78	4.400	1.168	
75	2.55	2.398	1.161	
75	3.78	2.200	1.273	
100	2.55	1.400	1.212	
100	3.78	1.199	1.369	

Program IML

Program IML allows us to calculate the width W and the effective dielectric constant ϵ_{eff} of the inverted microstrip line when the values of the following input data are known:

Z_0 = the characteristic impedance in ohms,
ϵ_r = the relative permittivity of the dielectric substrate,
$\mu_r = 1$,
d = the thickness of the dielectric substrate in m,
h = the thickness of the air gap in m.

Listing of Program IML

```
8700 REM IML
8702 CLS
8704 PRINT "INVERTED MICROSTRIP  LINE"
8706 PRINT
8708 PRINT "Data:"
8710 PRINT
8712 INPUT "Zo[ohm]=",ZO
8714 INPUT "er=",ER
8716 INPUT "d[m]=",D
8718 INPUT "h[m]=",H
8720 PRINT
8722 LET ZC=ZO
8724 LET PI=3.141593
8726 IF ZC <= 0 THEN  PRINT "Error: Zo <= 0": GOTO 8710
8728 IF ER <= 1 THEN  PRINT "Error: er <= 1": GOTO 8710
8730 IF D <= 0 THEN  PRINT "Error: d <= 0": GOTO 8710
8732 IF H <= 0 THEN  PRINT "Error: h <= 0": GOTO 8710
8734 PRINT "Please wait !"
8736 PRINT
8738 GOSUB 8750
8740 PRINT "Results, see Fig. 6.6"
8742 PRINT
8744 PRINT "ef="; FNE(U)* FNE(U)
8746 PRINT "W[m]=";U*H
8748 GOTO 8810
8750 REM Procedure IML
8752 DEF FNE(U)=1+Q*(C1-C2* LOG (U))*( SQR(ER)-1)
8754 DEF FNF(U)=6+(2* PI -6)* EXP (-(30.666/U)^.7528)
8756 DEF FNZ(U)=60/ FNE(U)* LOG ( FNF(U)/U+ SQR (1+(2/U)*(2/U))
)-ZC
8758 LET Q=D/H
8760 LET C1=(.5173-.1515* LOG (Q))*(.5173-.1515* LOG (Q))
8762 LET C2=(.3092-.1047* LOG (Q))*(.3092-.1047* LOG (Q))
8764 LET U=.2
8766 LET I=1
8768 LET DU=.2
8770 IF  FNZ(U)*FNZ(U+I*DU)<0 THEN  GOTO 8776
```

```
8772 LET I=I+1
8774 GOTO 8770
8776 LET U2=U+I*DU
8778 LET U1=U2-DU
8780 LET U3=U1+.381966*DU
8782 LET U4=U2-.381966*DU
8784 LET F4= ABS ( FNZ(U4))
8786 LET F3= ABS ( FNZ(U3))
8788 IF  ABS (U2-U1) <= .000001 THEN  LET U=(U1+U2)/2: RETURN
8790 IF F3>F4 THEN  GOTO 8800
8792 LET U2=U4
8794 LET U4=U3
8796 LET U3=U1+(U2-U1)*.381966
8798 GOTO 8786
8800 LET U1=U3
8802 LET U3=U4
8804 LET U4=U2-(U2-U1)*.381966
8806 LET F4= ABS ( FNZ(U4))
8808 GOTO 8788
8810 REM Chain subroutine
8812 PRINT
8814 INPUT"Enter 'CONT':",N$
8816 IF N$="CONT" THEN CLS:CHAIN "MENU",10
8818 GOTO 8812
```

6.7 COUPLED SLAB LINES

A transverse section of the parallel coupled slab lines is shown in Figure 6.7. The main electrical parameters of these lines are even-mode and odd-mode characteristic impedances denoted as Z_{0e} and Z_{0o}, respectively [17]. A knowledge of Z_{0e} and

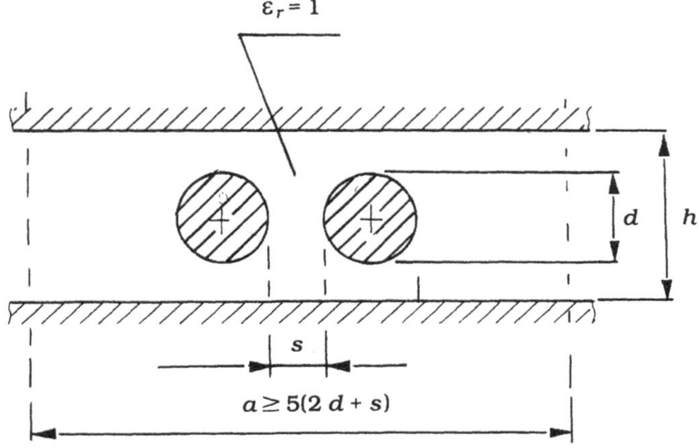

Figure 6.7 Cross section of coupled slab lines.

Z_{0o} as functions of the line structural parameters is essential for the design of many microwave devices such as different kinds of filters, directional couplers, *et cetera.* Hence, let us assume that the values of these impedances are known. Then designing the lines being considered consists of calculating $x = d/h$ and $y = s/h$, for instance, from the following set of equations [18]:

$$\begin{aligned} V_1(x,y) &= Z_{0e}(x,y) - Z_{0e} = 0 \\ V_2(x,y) &= Z_{0o}(x,y) - Z_{0o} = 0 \end{aligned} \tag{6.15}$$

where:

$$\begin{aligned}
Z_{0e}(x,y) &= 59.952 \ln\{0.523962/[f_1(x)f_2(x,y)f_3(x,y)]\} \\
Z_{0o}(x,y) &= 59.952 \ln\{0.523962 f_3(x,y)/[f_1(x)f_4(x,y)]\} \\
f_1(x) &= x\, a(x)/b(x) \\
f_2(x,y) &= \begin{cases} c(y) - xd(y) + e(x)g(y), & \text{for } y < 0.9 \\ 1 + 0.004 \exp(0.9 - y), & \text{for } y \geqslant 0.9 \end{cases} \\
f_3(x,y) &= \tanh[\pi(x + y)/2] \\
f_4(x,y) &= \begin{cases} k(y) - xl(y) + m(x)n(y), & \text{for } y < 0.9 \\ 1, & \text{for } y \geqslant 0.9 \end{cases} \\
a(x) &= 1 + \exp(16\, x - 18.272) \\
b(x) &= \sqrt{5.905 - x^4} \\
c(y) &= -\ 0.8107y^3 + 1.3401y^2 - 0.6929y + 1.0892 \\
&\quad + 0.014002/y - 0.000636/y^2 \\
d(y) &= 0.11 - 0.83y + 1.64y^2 - y^3 \\
e(x) &= -0.15 \exp(-13\, x) \\
g(y) &= 2.23 \exp(-7.01y + 10.24y^2 - 27.58y^3) \\
k(y) &= 1 + 0.01(-0.0726 - 0.2145/y + 0.222573/y^2 \\
&\quad - 0.012823/y^3) \\
l(y) &= 0.01(-0.26 + 0.6866/y + 0.0831/y^2 - 0.0076/y^3) \\
m(x) &= -0.1098 + 1.2138x - 2.2535x^2 + 1.1313x^3 \\
n(y) &= -0.019 - 0.016/y + 0.0362/y^2 - 0.00243/y^3
\end{aligned}$$

The algorithm for solving (6.15) should be accurate, reliable, rapidly convergent and simple. For solving these equations, a few different numerical methods have been tried. From the analysis performed, we see that the best results, in the sense of the above requirements, can be achieved by using the conventional Newton method. According to [3], the $(n + 1)$th approximation of the solution sought is found as follows:

$$x^{(n+1)} = x^{(n)} - \frac{1}{J}\left(V_1 \frac{\partial V_2}{\partial y} - V_2 \frac{\partial V_1}{\partial y}\right)$$
$$y^{(n+1)} = y^{(n)} + \frac{1}{J}\left(V_1 \frac{\partial V_2}{\partial x} - V_2 \frac{\partial V_1}{\partial x}\right) \tag{6.17}$$

where

$$J = \begin{bmatrix} \dfrac{\partial V_1}{\partial x}, \dfrac{\partial V_1}{\partial y} \\ \\ \dfrac{\partial V_2}{\partial x}, \dfrac{\partial V_2}{\partial y} \end{bmatrix} \neq 0 \tag{6.18}$$

and $(x^{(n)}, y^{(n)})$ is the n-th approximation, $n = 0, 1, 2, \ldots$.

We note that functions $V_1(x,y)$, $V_2(x,y)$ and their first partial derivatives are calculated at the point $(x^{(n)}, y^{(n)})$. The initial approximation of the solution can be evaluated from:

$$x^{(0)} = 4/\pi \exp[-Z_0/(59.952\sqrt{0.987 - 0.171k - 1.723k^2})]$$
$$y^{(0)} = 1/\pi \ln[(r + 1)/(r - 1)] - x^{(0)} \tag{6.19}$$

where

$$Z_0 = \sqrt{Z_{0e}Z_{0o}},$$
$$k = (Z_{0e} - Z_{0o})/(Z_{0e} + Z_{0o}),$$
$$r = [4/(\pi x^{(0)})]^{(0.001+1.117k-0.683k^2)}$$

We have found that the above initial approximation, entered into a computer program, significantly reduces the computation time and makes this program more reliable.

Formulas (6.15) and (6.16) together with (6.19) are used in the computer program CSL. In this program, the auxiliary function $U(x,y)$, defined as

$$U(x,y) = V_1^2(x,y) + V_2^2(x,y)$$

is also used. The search for the solution terminates if $U(x,y) \leqslant Z_{0e}Z_{0o}/1{,}000{,}000$, which ensures a relative accuracy of the approximation not worse than 0.1 percent for the given impedances Z_{0e} and Z_{0o}. In program CSL, apart from computation statements, there are some instructions for protection from different kind of errors.

For instance, the calculations will not be made if $Z_{0e} \leqslant Z_{0o}$, and then the appropriate report is printed. Similarly, the Jacobian given by (6.18) should assume nonzero values during all iterations. Errors due to dividing by zero, although they are of small probability, are eliminated by means of instructions written in the program.

As has been confirmed, (6.16) ensures a very good agreement of calculations with the accurate numerical results for a wide range of geometrical dimensions. For $0.1 \leqslant d/h \leqslant 0.8$ and $0.1 \leqslant s/h$, the maximum relative deviation between the accurate numerical results given in [19] and the interpolated values is not greater than 0.27 percent for the impedance Z_{0e}. Similarly, for $0.1 \leqslant d/h \leqslant 0.8$ and $0.1 \leqslant s/h$, the maximum relative deviation evaluated for the impedance Z_{0o} is equal to 0.65 percent.

Check Calculations

Some check results computed with the program CSL are given in Table 6.7.

Table 6.7
Geometrical Dimensions of Coupled Slab Lines as a Function of Their Even- and Odd-Mode Characteristic Impedances

		$\epsilon_r = 1, \mu_r = 1$	
Z_{0e}, Ω	Z_{0o}, Ω	d/h	s/h
51.607	48.443	0.5483	0.7882
52.895	47.263	0.5460	0.6119
55.277	45.227	0.5418	0.4439
59.845	41.774	0.5286	0.2802
69.371	36.038	0.4893	0.1460

Program CSL

The computer program CSL allows us to calculate the geometrical dimensions of the coupled slab lines for the following input data:

Z_{0e} = the even-mode characteristic impedance in ohms,
Z_{0o} = the odd-mode characteristic impedance in ohms,
h = the line thickness in m.

As output results, we obtain the diameter *d* of the rods and the width *s* of the slot, both expressed in m.

Listing of Program CSL

```
8900 REM CSL
8902 CLS
8904 PRINT "COUPLED SLAB LINES"
8906 PRINT
8908 PRINT "Data:"
8910 PRINT
8912 INPUT "Zoe[ohm]=",ZOE
8914 INPUT "Zoo[ohm]=",ZOO
8916 INPUT "h[m]=",TH
8918 PRINT "er=1"
8920 PRINT "ur=1"
8922 PRINT
8924 IF TH <= 0 THEN  PRINT "Error:h <= 0": GOTO 8910
8926 IF ZOE <= 0 THEN  PRINT "Error:Zoe <= 0": GOTO 8910
8928 IF ZOO <= 0 THEN  PRINT "Error:Zoo <= 0": GOTO 8910
8930 IF ZOE<ZOO THEN  PRINT "Error: Zoe<Zoo": GOTO 8910
8932 IF ZOE=ZOO THEN  PRINT "The lines are not coupled.": GOTO
9068
8934 PRINT "Please wait !"
8936 PRINT
8938 LET PI=3.141593
8940 LET Z= SQR (ZOE*ZOO)
8942 LET K=(ZOE-ZOO)/(ZOE+ZOO)
8944 LET XO=4/ PI * EXP (-Z/(59.952* SQR(ABS(.987-.171*K-1.723*
K*K))))
8946 LET R=(4/( PI *XO))^(.001+1.117*K-.683*K*K)
8948 LET YO= ABS (1/ PI * LOG ((R+1)/(R-1))-XO)
8950 LET X=XO
8952 LET Y=YO
8954 GOSUB 8966
8956 PRINT "Results, see Fig. 6.7"
8958 PRINT
8960 PRINT "d[m]=";X*TH
8962 PRINT "s[m]=";Y*TH
8964 GOTO 9068
8966 REM Procedure CSL
8968 GOSUB 9024
8970 LET UO=U
8972 LET VO=V
8974 IF (UO*UO+VO*VO)<Z*Z/1000000! THEN  RETURN
8976 LET DH=.0001
8978 LET X=X+DH
8980 GOSUB 9024
8982 LET U1=U
8984 LET V1=V
8986 LET X=X-DH
8988 LET Y=Y+DH
```

```
8990 GOSUB 9024
8992 LET U2=U
8994 LET V2=V
8996 LET D1=(U1-UO)/DH
8998 LET D2=(U2-UO)/DH
9000 LET D3=(V1-VO)/DH
9002 LET D4=(V2-VO)/DH
9004 LET DET=D1*D4-D2*D3
9006 IF  ABS (DET)>1E-09 THEN  GOTO 9012
9008 LET X=1.01*X
9010 GOTO 8968
9012 LET X= ABS (X-(UO*D4-VO*D2)/DET)
9014 LET Y= ABS (Y+(UO*D3-VO*D1)/DET)
9016 IF Y<9.000001E-02 THEN  PRINT "s/h<0.1": GOTO 9068
9018 IF X<9.000001E-02 THEN  PRINT "d/h<0.1": GOTO 9068
9020 IF X>.81 THEN  PRINT "d/h>0.8": GOTO 9068
9022 GOTO 8968
9024 REM Subroutine 9024
9026 LET A=1+ EXP (16*X-18.272)
9028 LET B= SQR (5.905-X*X*X*X)
9030 LET H= EXP ( PI *(X+Y)/2)
9032 LET F1=X*A/B
9034 LET F3=(H*H-1)/(H*H+1)
9036 IF Y >= .9 THEN  LET F2=1+.004* EXP (.9-Y): LET F4=1: GOTO
 9058
9038 LET C=-.8107*Y^3+1.3401*Y^2-.6929*Y+1.0892+.014002/Y-.0006
36/Y^2
9040 LET D=.11-.83*Y+1.64*Y^2-Y^3
9042 LET E=-.15* EXP (-13*X)
9044 LET G=2.23* EXP (-7.01*Y+10.24*Y^2-27.58*Y^3)
9046 LET K=1+.01*(-.0726-.2145/Y+.222573/Y^2-.012823/Y^3)
9048 LET L=.01*(-.26+.6866/Y+.0831/Y^2-.0076/Y^3)
9050 LET M=-.1098+1.2138*X-2.2535*X^2+1.1313*X^3
9052 LET N=-.019-.016/Y+.0362/Y^2-.00243/Y^3
9054 LET F2=C-X*D+E*G
9056 LET F4=K-X*L+M*N
9058 LET ZE=59.952* LOG( ABS (.5239621/(F1*F2*F3)))
9060 LET ZO=59.952* LOG( ABS (.5239621*F3/(F1*F4)))
9062 LET U=ZE-ZOE
9064 LET V=ZO-ZOO
9066 RETURN
9068 REM Chain subroutine
9070 PRINT
9072 INPUT"Enter 'CONT':",N$
9074 IF N$="CONT" THEN CLS:CHAIN "MENU",10
9076 GOTO 9070
```

6.8 EDGE COUPLED TRIPLATE STRIPLINES

A transverse section of the edge coupled triplate striplines is shown in Figure 6.8. As Gunston pointed out in [2], the strip thicknesses used in microwave printed circuits are, most frequently, sufficiently small to allow quite accurate designing of circuit components by means of the zero-thickness formulas given in [20]:

$$Z_{0e}(W, S) = 29.976\pi \sqrt{\frac{\mu_r}{\epsilon_r} \frac{K'(k_e)}{K(k_e)}}, \Omega$$
$$Z_{0o}(W, S) = 29.976\pi \sqrt{\frac{\mu_r}{\epsilon_r} \frac{K'(k_o)}{K(k_o)}}, \Omega \tag{6.20}$$

where:

k_e = $\tanh[\pi W/(2b)] \tanh[\pi(W + S)/(2b)]$,
k_o = $\tanh[\pi W/(2b)] \coth[\pi(W + S)/(2b)]$,
ϵ_r = the relative permittivity of the dielectric substrate,
μ_r = the relative permeability of the dielectric substrate,
$K(k)$ = the complete elliptic integral of the first kind,
$K'(k)$ = the complementary to $K(k)$ complete elliptic integral of the first kind, see Appendix 1.

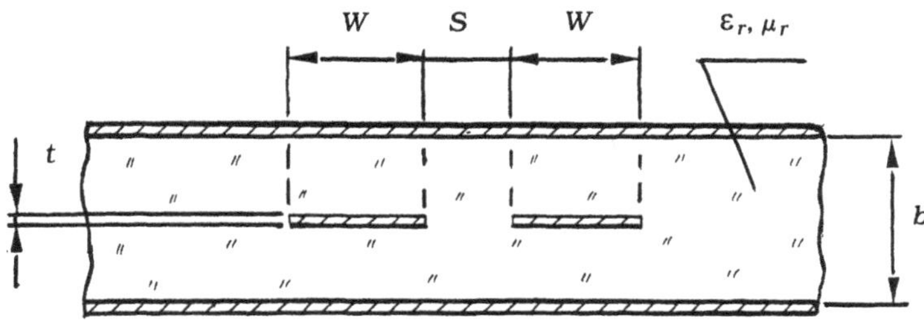

Figure 6.8 Cross section of edge coupled triplate striplines.

The wavelength λ_f within the lines is related to the free space wavelength λ_0 as follows:

$$\lambda_f = \lambda_0/\sqrt{\epsilon_r \mu_r}$$

The synthesis of the lines being considered (i.e., evaluation of their geometrical dimensions), for given values of impedances Z_{0e} and Z_{0o}, can be done by using the following algorithm. In the first step, we calculate the values of the modular constants:

$$q_e = \exp[-\pi K'(k_e)/K(k_e)]$$
$$q_o = \exp[-\pi K'(k_o)/K(k_o)] \tag{6.21}$$

Next, the moduli k_e and k_o are calculated by using (6.6) or (6.8), respectively. Finally, the widths W and S of strips and of the slot, respectively, are calculated from the formulas:

$$W = \frac{2b}{\pi}\tanh^{-1}\sqrt{k_e k_o}$$
$$S = \frac{2b}{\pi}\tanh^{-1}\sqrt{k_e/k_o} - W \tag{6.22}$$

where b is the dielectric substrate thickness.

For some applications, the assumption of the zero strip thickness is not valid. In these cases, we can utilize the following approximative expressions [6,20]:

$$Z_{0e}(W, S, t) = \frac{30\pi(b - t)}{\sqrt{\epsilon_r}(W + A_e b C_f)}, \Omega$$
$$Z_{0o}(W, S, t) = \frac{30\pi(b - t)}{\sqrt{\epsilon_r}(W + A_o b C_f)}, \Omega \tag{6.23}$$

with

$$A_e = [\ln 2 + \ln(1 + \tanh\theta)]/(2\pi \ln 2)$$
$$A_o = [\ln 2 + \ln(1 + \coth\theta)]/(2\pi \ln 2)$$
$$\theta = \pi S/(2b)$$
$$C_f = 2\ln\left(\frac{2b - t}{b - t}\right) - \frac{t}{b}\ln\left[\frac{t(2b - t)}{(b - t)^2}\right]$$

The accuracy of these formulas is the best for $t/b \leqslant 0.1$ and $W/b \geqslant 0.35$ [6].

Formulas (6.20) to (6.23) constitute the mathematical basis for the computer program CTL, presented below. The coupled triplate striplines with strip thickness $t \leqslant 0.01b$ (see Figure 6.8) are designed by using the zero-thickness formulas (6.20). If $t > 0.01b$, the approximate formulas (6.23) are used for the design.

Check Calculations

Some typical results of calculations obtained by using the program CTL are presented in Tables 6.8 and 6.9.

Table 6.8
Geometrical Dimensions of Edge Coupled Triplate Striplines as a Function of Their Even- and Odd-mode Characteristic Impedances

		$b = 2\ 10^{-3}\ m,\ t = 0\ m,\ \mu_r = 1$		
Z_{0e}, Ω	Z_{0o}, Ω	ϵ_r	W, m	S, m
51.60	48.44	2.55	$1.475 \cdot 10^{-3}$	$1.365 \cdot 10^{-3}$
55.27	45.23	2.55	$1.446 \cdot 10^{-3}$	$0.655 \cdot 10^{-3}$
69.37	36.04	2.55	$1.208 \cdot 10^{-3}$	$0.106 \cdot 10^{-3}$
86.74	28.82	2.55	$0.910 \cdot 10^{-3}$	$0.164 \cdot 10^{-4}$
51.60	48.44	9.60	$0.386 \cdot 10^{-3}$	$1.679 \cdot 10^{-3}$
55.27	45.23	9.60	$0.381 \cdot 10^{-3}$	$0.952 \cdot 10^{-3}$
69.37	36.04	9.60	$0.325 \cdot 10^{-3}$	$0.279 \cdot 10^{-3}$
86.74	28.82	9.60	$0.236 \cdot 10^{-3}$	$0.831 \cdot 10^{-4}$

Table 6.9
Geometrical Dimensions of Finite-Thickness Edge Coupled Striplines as a Function of Their Even- and Odd-Mode Characteristic Impedances

		$b = 2\ 10^{-3}\ m,\ t = 3\ 10^{-5}\ m,\ \mu_r = 1$		
Z_{0e}, Ω	Z_{0o}, Ω	ϵ_r	W, m	S, m
51.60	48.44	2.55	$1.397 \cdot 10^{-3}$	$1.407 \cdot 10^{-3}$
55.27	45.23	2.55	$1.372 \cdot 10^{-3}$	$0.694 \cdot 10^{-3}$
69.37	36.04	2.55	$1.151 \cdot 10^{-3}$	$0.126 \cdot 10^{-3}$
86.74	28.82	2.55	$0.866 \cdot 10^{-3}$	$0.226 \cdot 10^{-4}$

Program CTL

The program CTL makes it possible to design the coupled triplate striplines when their following parameters are known:

Z_{0e} = the even-mode characteristic impedance in ohms,
Z_{0o} = the odd-mode characteristic impedance in ohms,
ϵ_r = the relative permittivity of the dielectric substrate,
$\mu_r = 1$,
h = the thickness of the dielectric substrate in m,
t = the thickness of strips in m.

In the design process, the widths W of strip and S of the slot are calculated.

Listing of Program CTL

```
9300 REM CTL
9302 CLS:DEFDBL A-Z:DEFINT I
9304 PRINT "COUPLED TRIPLATE STRIPLINES"
9306 PRINT
9308 PRINT "Data:"
9310 PRINT
9312 INPUT "Zoe[ohm]=",ZOE
9314 INPUT "Zoo[ohm]=",ZOO
9316 INPUT "er=",ER
9318 PRINT "ur=1"
9320 INPUT "b[m]=",B
9322 INPUT "t[m]=",T
9324 PRINT
9326 IF ZOE <= 0 THEN  PRINT "Error:Zoe <= 0": GOTO 9310
9328 IF ZOO <= 0 THEN  PRINT "Error:Zoo <= 0": GOTO 9310
9330 IF ER<1 THEN  PRINT "Error:er<1": GOTO 9310
9332 IF B <= 0 THEN  PRINT "Error:b <= 0": GOTO 9310
9334 IF T<0 THEN  PRINT "t<0": GOTO 9310
9336 IF ZOO >= ZOE THEN  PRINT "Zoo >= Zoe": GOTO 9310
9338 IF T>B/10 THEN  PRINT "t>b/10": GOTO 9310
9340 PRINT "Please wait !"
9342 PRINT
9344 LET ZCE=ZOE
9346 LET ZCO=ZOO
9348 GOSUB 9360
9350 PRINT "Results, see Fig. 6.8"
9352 PRINT
9354 PRINT USING "W[m]=##.#######";W
9356 PRINT USING "S[m]=##.#######";S
9358 GOTO 9478
9360 REM Procedure CTL
9362 LET Z=ZCE
9364 LET PI=3.141592653#
9366 GOSUB 9440
9368 LET KE=K
9370 LET Z=ZCO
9372 GOSUB 9440
9374 LET KO=K
9376 LET CC= SQR (KE*KO)
9378 LET W=B/ PI * LOG((1+CC)/(1-CC))
9380 LET CC= SQR (KE/KO)
9382 LET S=B/ PI * LOG((1+CC)/(1-CC))-W
9384 IF T <= B/100 THEN  RETURN
9386 LET CF=2* LOG((2*B-T)/(B-T))-T* LOG(T*(2*B-T)/(B-T)^2)/B
9388 LET CC=2* PI * LOG(2)
9390 GOSUB 9460
9392 LET UO=U
9394 LET VO=V
9396 IF (UO*UO+VO*VO) <= ZCE*ZCO/1000000! THEN  RETURN
9398 LET DH=.0000001
9400 LET W=W+DH
9402 GOSUB 9460
```

```
9404 LET U1=U
9406 LET V1=V
9408 LET W=W-DH
9410 LET S=S+DH
9412 GOSUB 9460
9414 LET U2=U
9416 LET V2=V
9418 LET D1=(U1-UO)/DH
9420 LET D2=(U2-UO)/DH
9422 LET D3=(V1-VO)/DH
9424 LET D4=(V2-VO)/DH
9426 LET DET=D1*D4-D2*D3
9428 IF  ABS (DET) >= 1E-09 THEN  GOTO 9434
9430 LET W=1.01*W
9432 GOTO 9390
9434 LET W= ABS (W-(UO*D4-VO*D2)/DET)
9436 LET S= ABS (S+(UO*D3-VO*D1)/DET)
9438 GOTO 9390
9440 REM Subroutine k(q)
9442 LET Q= EXP (-Z/29.976* SQR (ER))
9444 LET L=0
9446 LET M=.5
9448 FOR I=1 TO 6
9450 LET L=L+Q^(I*(I-1))
9452 LET M=M+Q^(I*I)
9454 NEXT I
9456 LET K= SQR (Q)*(L/M)*(L/M)
9458 RETURN
9460 REM Subroutine 9460
9462 LET O= PI *S/(2*B)
9464 LET P= EXP (O)
9466 LET TH=(P-1/P)/(P+1/P)
9468 LET AE=( LOG(2)+ LOG(1+TH))/CC
9470 LET AO=( LOG(2)+ LOG(1+1/TH))/CC
9472 LET U=30* PI *(B-T)/( SQR (ER)*(W+AE*B*CF))-ZCE
9474 LET V=30* PI *(B-T)/( SQR (ER)*(W+AO*B*CF))-ZCO
9476 RETURN
9478 REM Chain subroutine
9480 PRINT
9482 INPUT"Enter 'CONT':",N$
9484 IF N$="CONT" THEN CLS:CHAIN "MENU",10
9486 GOTO 9480
```

6.9 COUPLED MICROSTRIP LINES

The basic structure of coupled microstrip lines is presented in Figure 6.9. These lines are widely employed in numerous microwave integrated circuits, such as coupled transmission line directional couplers, bandpass filters, and dc blocks. Therefore, their design has been extensively studied for many years. Consequently, different semianalytic expressions, numerical data, and closed-form formulas for design have been published in the literature [2,6,13,22]. From the analysis per-

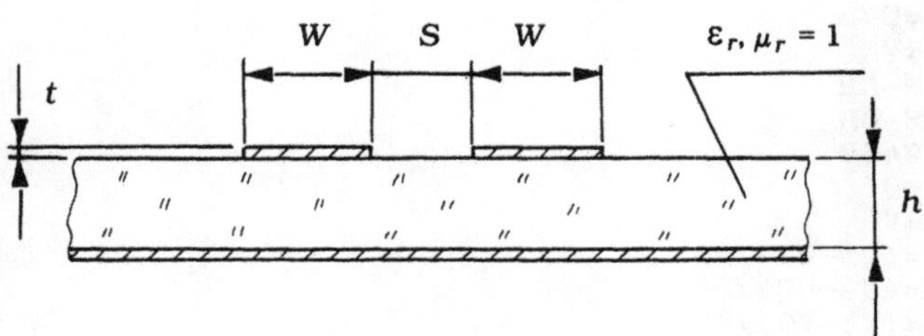

Figure 6.9 Cross section of coupled microstrip lines.

formed, we see that the closed-form expressions given in [22] are most useful for computer-aided design. For this reason, some of them are utilized in the design algorithm presented below. This algorithm makes possible the calculation of the following parameters of the lines under analysis:

$Z_{0e}(0)$ = the static even-mode characteristic impedance,
$Z_{0o}(0)$ = the static odd-mode characteristic impedance,
$\epsilon_{\text{eff}_e}(f)$ = the frequency dependent even-mode effective dielectric constant,
$\epsilon_{\text{eff}_o}(f)$ = the frequency dependent odd-mode effective dielectric constant.

Here, we note that in all the characteristics considered above, we did not take into account the finite thickness of the strip conductors. In spite of the assumed simplifications, these expressions ensure good accuracy of calculations for effective dielectric constants, as well as for the characteristic impedances. According to [22], the accuracy of calculations is better than 1% for the above-mentioned parameters if $0.1 \leqslant W/h \leqslant 10$, $1 \leqslant \epsilon_r \leqslant 10$, $t \leqslant 0.01h$ and $f \leqslant 6$ GHz.

Let us assume that $u = W/h$ denotes the normalized strip width and $g = S/h$ is the normalized line spacing. In this notation, the static even-mode effective dielectric constant is

$$\epsilon_{\text{eff}_e}(0) = \frac{\epsilon_r + 1}{2} + \frac{\epsilon_r - 1}{2}\left(1 + \frac{10}{v}\right)^{-a_e(v)b_e(\epsilon_r)} \tag{6.24}$$

where

$$v = u(20 + g^2)/(10 + g^2) + g\exp(-g)$$

$$a_e(v) = 1 + \frac{1}{49}\ln\left[\frac{v^4 + (v/52)^2}{v^4 + 0.432}\right] + \frac{1}{18.7}\ln\left[1 + \left(\frac{v}{18.1}\right)^3\right]$$

$$b_e(\epsilon_r) = 0.564[(\epsilon_r - 0.9)/(\epsilon_r + 3)]^{0.053}$$

Similarly, the static odd-mode effective dielectric constant is

$$\epsilon_{\text{eff}_o}(0) = \left[\frac{\epsilon_r + 1}{2} + a_o(u, \epsilon_r) - \epsilon_{\text{eff}}(0)\right] \exp(-c_o g^{d_o}) + \epsilon_{\text{eff}}(0) \tag{6.25}$$

where

$$a_o(u, \epsilon_r) = 0.7287 \left[\epsilon_{\text{eff}}(0) - \frac{\epsilon_r + 1}{2}\right] [1 - \exp(-0.179u)]$$
$$b_o(\epsilon_r) = 0.747\epsilon_r/(0.15 + \epsilon_r)$$
$$c_o = b_o(\epsilon_r) - [b_o(\epsilon_r) - 0.207] \exp(-0.414u)$$
$$d_o = 0.593 + 0.694 \cdot \exp(-0.562u)$$

$\epsilon_{\text{eff}}(0)$ refers to static zero-thickness single microstrip of width W (see (6.12)).

The frequency dispersion is provided for both modes in the same form, namely:

$$\epsilon_{\text{eff}_{e,o}}(f_n) = \epsilon_r - \frac{\epsilon_r - \epsilon_{\text{eff}_{e,o}}(0)}{1 + F_{e,o}(f_n)} \tag{6.26}$$

where

$$f_n = (f/\text{GHz})(\text{h/mm})$$

The functions $F_e(f_n)$ and $F_o(f_n)$ are given by the following expressions:

$$F_e(f_n) = P_1 P_2 [(P_3 P_4 + 0.1844 P_7) f_n]^{1.5763} \tag{6.27}$$
$$F_o(f_n) = P_1 P_2 [(P_3 P_4 + 0.1844) f_n P_{15}]^{1.5763}$$

with

$$P_1 = 0.27488 + [0.6315 + 0.525/(1 + 0.0157 f_n)^{20}]u - 0.065683 \exp(-8.7513u)$$
$$P_2 = 0.33622[1 - \exp(-0.03442\epsilon_r)]$$
$$P_3 = 0.0363 \exp(-4.6u)\{1 - \exp[-(f_n/38.7)^{4.97}]\}$$
$$P_4 = 1 + 2.751\{1 - \exp[-(\epsilon_r/15.916)^8]\}$$
$$P_5 = 0.334 \exp[-3.3(\epsilon_r/15)^3] + 0.746$$

$$P_6 = P_5 \exp[-(f_n/18)^{0.368}]$$
$$P_7 = 1 + 4.069 P_6 g^{0.479} \exp(-1.347 g^{0.595} - 0.17 g^{2.5})$$
$$P_8 = 0.7168\{1 + 1.076/[1 + 0.0576(\epsilon_r - 1)]\}$$
$$P_9 = P_8 - 0.7913\{1 - \exp[-(f_n/20)^{1.424}]\} \arctan[2.481(\epsilon_r/8)^{0.946}]$$
$$P_{10} = 0.242(\epsilon_r - 1)^{0.55}$$
$$P_{11} = 0.6366[\exp(-0.3401 f_n) - 1] \arctan[1.263(u/3)^{1.629}]$$
$$P_{12} = P_9 + (1 - P_9)/(1 + 1.183 u^{1.376})$$
$$P_{13} = 1.695 P_{10}/(0.414 + 1.605 P_{10})$$
$$P_{14} = 0.8928 + 0.1072\{1 - \exp[-0.42(f_n/20)^{3.215}]\}$$
$$P_{15} = |1 - 0.8928(1 + P_{11}) P_{12} \exp(-P_{13} g^{1.092})/P_{14}|$$

The static even-mode characteristic impedance $Z_{0e}(0)$ is expressed as:

$$Z_{0e}(0) = Z_L(0) \frac{\sqrt{\epsilon_{\mathrm{eff}}(0)/\epsilon_{\mathrm{eff}_e}(0)}}{1 - [Z_L(0)/377]\sqrt{\epsilon_{\mathrm{eff}}(0)}\, Q_4} \tag{6.28}$$

where $Z_L(0)$ denotes the characteristic impedance of the single microstrip line [13] of width W and

$$Q_1 = 0.8695 u^{0.194}$$
$$Q_2 = 1 + 0.7519 g + 0.189 g^{2.31}$$
$$Q_3 = 0.1975 + [16.6 + (8.4/g)^6]^{-0.387} + \ln\{g^{10}/[1 + (g/3.4)^{10}]\}/241$$
$$Q_4 = \frac{2Q_1/Q_2}{\exp(-g) u^{Q_3} + [2 - \exp(-g)] u^{-Q_3}}$$

Similarly, the static odd-mode characteristic impedance $Z_{0o}(0)$ is:

$$Z_{0o}(0) = Z_L(0) \frac{\sqrt{\epsilon_{\mathrm{eff}}(0)/\epsilon_{\mathrm{eff}_o}(0)}}{1 - [Z_L(0)/377]\sqrt{\epsilon_{\mathrm{eff}}(0)}\, Q_{10}} \tag{6.29}$$

with

$$Q_5 = 1.794 + 1.14 \ln[1 + 0.638/(g + 0.517 g^{2.43})]$$

$$Q_6 = 0.2305 + \ln\{g^{10}/[1 + (g/5.8)^{10}]\}/281.3 + \ln(1 + 0.598\, g^{1.154})/5.1$$
$$Q_7 = (10 + 190g^2)/(1 + 82.3g^3)$$
$$Q_8 = \exp[-6.5 - 0.95\ln(g) - (g/0.15)^5]$$
$$Q_9 = \ln(Q_7)(Q_8 + 1/16.5)$$
$$Q_{10} = \{Q_2Q_4 - Q_5\exp[\ln(u)Q_6u^{-Q_9}]\}/Q_2$$

These formulas make possible the calculation of the characteristic impedances $Z_{0e}(u, g, 0)$ and $Z_{0o}(u, g, 0)$, provided that we know the geometric dimensions and the relative permittivity ϵ_r of the dielectric substrate. In practice, however, usually the reverse problem is solved, which is equivalent to solving two nonlinear equations where the width of strips and the width of the line spacing are the unknown variables. Hence, let us assume that the values of Z_{0e}, Z_{0o}, ϵ_r and frequency f are known. Then, the design consists of solving the following set of equations:

$$\begin{aligned} V_1(u, g) &= Z_{0e}(u, g, 0) - Z_{0e} = 0 \\ V_2(u, g) &= Z_{0o}(u, g, 0) - Z_{0o} = 0 \end{aligned} \tag{6.30}$$

The problem (6.30) can also be reformulated to the corresponding minimization, for instance:

$$\min_{(u,\, g)} \quad U(u, g) = V_1^2(u, g) + V_2^2(u, g) \tag{6.31}$$

From (6.30) and (6.31), function $U(u, g)$ hence assumes its global minimum equal to zero at point (u, g) being sought. The problem (6.31) can be solved by means of different methods of minimization (e.g., the steepest descent, Fletcher-Reeves or Davidon-Fletcher-Powell methods).

Generally, the numerical algorithm for solving the set of equations (6.30) or minimization problem (6.31) should be accurate enough, numerically reliable, rapidly convergent, and simple. From the analysis performed, we see that the best results, in the sense of the above requirements, can be achieved by using the conventional Newton method [3], the outline of which is also presented in Section 6.7.

The initial approximation of the solution being sought is found from [23]

$$\begin{aligned} u^{(0)} &= |F_1(Z_0, \epsilon_r) \cdot F_2(k)| \\ g^{(0)} &= |F_1(Z_0, \epsilon_r) \cdot F_3(k, \epsilon_r)| \end{aligned} \tag{6.32}$$

where

$$Z_0 = \sqrt{Z_{0e}Z_{0o}}$$
$$k = (Z_{0e} - Z_{0o})/(Z_{0e} + Z_{0o})$$
$$F_1(Z_0, \epsilon_r) = 8\sqrt{A(7 + 4/\epsilon_r)/11 + (1 + 1/\epsilon_r)/0.81}/A$$
$$A = \exp(Z_0\sqrt{\epsilon_r + 1}/42.4) - 1$$
$$F_2(k) = \sum_{i=1}^{6} a_i k^{(i-1)}$$
$$F_3(k, \epsilon_r) = \sum_{i=1}^{6} [b_i - c_i(9.6 - \epsilon_r)](0.6 - k)^{(i-1)}$$

for $k \leqslant 0.5$ and $F_3(k,\epsilon_r) = 2(1 - k)F_3(k = 0.5, \epsilon_r)$ for $k > 0.5$.

The values of coefficients a_i, b_i and c_i, where $i = 1, 2, \ldots, 6$, are given in Table 6.10. Here we note that in the CML program presented below, function $U(u, g)$ is used as a basis of the stop criterion. The search of the solution terminates if $U(u, g)$ is less than $Z_{0e}Z_{0o}/1{,}000{,}000$.

Table 6.10
Coefficients of the Approximating Functions $F_2(k)$ and $F_3(k, \epsilon_r)$ given in Section 6.9

i	a_i	b_i	c_i
1	1	0.020	0.002
2	− 0.301	− 0.623	− 0.347
3	3.209	17.192	7.171
4	− 27.282	− 68.946	− 36.910
5	56.609	104.740	76.132
6	− 37.746	− 16.148	− 51.616

Check Calculations

Some typical results of calculations, obtained by means of program CML, are given in Tables 6.11 and 6.12.

Table 6.11
Some Results of Design of Coupled Microstrip Lines

		$h = 10^{-3}$ m, $\epsilon_r = 2.55$, $f = 6\ 10^9$ Hz			
Z_{0e}, Ω	Z_{0o}, Ω	W, m	S, m	ϵ_{eff_e}	ϵ_{eff_o}
55.27	45.23	$2.763 \cdot 10^{-3}$	$1.213 \cdot 10^{-3}$	2.261	2.009
69.37	36.04	$2.314 \cdot 10^{-3}$	$0.123 \cdot 10^{-3}$	2.239	1.882
86.74	28.82	$1.744 \cdot 10^{-3}$	$0.988 \cdot 10^{-5}$	2.180	1.833

Table 6.12
Some Results of Design of Coupled Microstrip Lines

		$h = 10^{-3}\ m,\ \epsilon_r = 10.20, f = 6\ 10^9\ Hz$			
Z_{0e}, Ω	Z_{0o}, Ω	W, m	S, m	ϵ_{eff_e}	ϵ_{eff_o}
55.27	45.23	$0.922 \cdot 10^{-3}$	$1.336 \cdot 10^{-3}$	7.733	6.382
69.37	36.04	$0.785 \cdot 10^{-3}$	$0.299 \cdot 10^{-3}$	7.695	5.846
86.74	28.82	$0.581 \cdot 10^{-3}$	$0.711 \cdot 10^{-4}$	7.342	5.692

Program CML

In order to calculate the geometrical dimensions of coupled microstrip lines, we must introduce into the computer memory values of the following quantities:

Z_{0e} — the even-mode characteristic impedance in ohms,
Z_{0o} — the odd-mode characteristic impedance in ohms,
ϵ_r — the relative permittivity of the dielectric substrate,
$\mu_r = 1$
h — the thickness of the dielectric substrate in m,
f — the signal frequency in Hz.

As output results we obtain the values of W, S, $\epsilon_{eff_e}(f)$ and $\epsilon_{eff_o}(f)$ (see Figure 6.9).

Listing of Program CML

```
9600 REM CML
9602 CLS:DEFDBL A-Z:DEFINT I
9604 DIM A(6)
9606 DIM B(6)
9608 DIM C(6)
9610 PRINT "COUPLED MICROSTRIP LINES"
9612 PRINT
9614 PRINT "Data:"
9616 PRINT
9618 INPUT "Zoe[ohm]=",ZOE
9620 INPUT "Zoo[ohm]=",ZOO
9622 INPUT "er=",ER
9624 PRINT "ur=1"
9626 INPUT "h[m]=",H
9628 INPUT "f[Hz]=",F
9630 PRINT
9632 IF ZOE <= 0 THEN  PRINT "Error:Zoe <= 0": GOTO 9616
9634 IF ZOO <= 0 THEN  PRINT "Error:Zoo <= 0": GOTO 9616
9636 IF ZOE=ZOO THEN  PRINT "The lines are not coupled !": GOTO
 9868
```

```
9638 IF ZOE<ZOO THEN  PRINT "Error:Zoe<Zoo": GOTO 9616
9640 IF ER<1 THEN  PRINT "Error:er<1": GOTO 9616
9642 IF H <= 0 THEN  PRINT "Error:h <= 0": GOTO 9616
9644 IF F<0 THEN PRINT "Error:f<0": GOTO 9616
9646 PRINT "Please wait !"
9648 PRINT
9650 LET ZCE=ZOE
9652 LET ZCO=ZOO
9654 GOSUB 9670
9656 PRINT "Results, see Fig. 6.9"
9658 PRINT
9660 PRINT USING"W[m]=##.######";U*H
9662 PRINT USING"S[m]=##.######";G*H
9664 PRINT USING"eeffe=##.######";EFEF
9666 PRINT USING"eeffo=##.######";EFOF
9668 GOTO 9868
9670 REM Procedure CML
9672 LET PI=3.141592653#
9674 FOR I=1 TO 6
9676 READ A(I)
9678 READ B(I)
9680 READ C(I)
9682 NEXT I
9684 DATA 1,.02,.002,-.301,-.623,-.347,3.209,17.192,7.171,-27.2
82,-68.946,-36.91,56.609,104.74,76.132,-37.746,-16.148,-51.616
9686 LET ZO= SQR (ZCE*ZCO)
9688 LET K=(ZCE-ZCO)/(ZCE+ZCO)
9690 LET AW= EXP (ZO/42.4* SQR (ER+1))-1
9692 LET UW=8/AW* SQR (AW/11*(7+4/ER)+1/.81*(1+1/ER))
9694 LET F2=0
9696 LET F3=0
9698 FOR I=1 TO 6
9700 LET F2=F2+A(I)*K^(I-1)
9702 NEXT I
9704 IF K >= .5 THEN  LET KK=.5: GOTO 9708
9706 LET KK=K
9708 FOR I=1 TO 6
9710 LET F3=F3+(B(I)-C(I)*(9.600001-ER))*(.6-KK)^(I-1)
9712 NEXT I
9714 IF K >= .5 THEN  LET F3=2*F3*(1-K)
9716 LET U= ABS (UW*F2)
9718 LET G= ABS (UW*F3)
9720 LET ST=.00001
9722 GOSUB 9774
9724 LET MO=ZE
9726 LET NO=ZO
9728 LET U=U+ST
9730 GOSUB 9774
9732 LET M1=ZE
9734 LET N1=ZO
9736 LET U=U-ST
9738 LET G=G+ST
9740 GOSUB 9774
9742 LET M2=ZE
9744 LET N2=ZO
```

```
9746 LET D1=(M1-MO)/ST
9748 LET D2=(M2-MO)/ST
9750 LET D3=(N1-NO)/ST
9752 LET D4=(N2-NO)/ST
9754 LET DET=D1*D4-D2*D3
9756 IF  ABS (DET)>1E-09 THEN  GOTO 9762
9758 LET U=1.01*U
9760 GOTO 9722
9762 LET U= ABS (U-(MO*D4-NO*D2)/DET)
9764 LET G= ABS (G+(MO*D3-NO*D1)/DET)
9766 GOSUB 9774
9768 IF (ZE*ZE+ZO*ZO) <= ZCE*ZCO/1000000! THEN  RETURN
9770 LET ST=.9*ST
9772 GOTO 9722
9774 REM Subroutine 9774
9776 LET V=U*(20+G*G)/(10+G*G)+G* EXP (-G)
9778 LET AE=1+1/49* LOG((V^4+(V/52)^2)/(V^4+.432))+1/18.7* LOG(
1+(V/18.1)^3)
9780 LET BE=.564*((ER-.9)/(ER+3))^.053
9782 LET EFEO=(ER+1)/2+(ER-1)/2*(1+10/V)^(-AE*BE)
9784 LET AU=1+1/49* LOG((U^4+(U/52)^2)/(U^4+.432))+1/18.7* LOG(
1+(U/18.1)^3)
9786 LET EF=(ER+1)/2+(ER-1)/2*(1+10/U)^(-AU*BE)
9788 LET AO=.7287*(EF-(ER+1)/2)*(1- EXP (-.179*U))
9790 LET BO=.747*ER/(.15+ER)
9792 LET CO=BO-(BO-.207)* EXP (-.414*U)
9794 LET DO=.593+.694* EXP (-.562*U)
9796 LET EFOO=((ER+1)/2+AO-EF)* EXP (-CO*G^DO)+EF
9798 LET CN=F*H*.000001
9800 LET P1=.27488+(.6315+.525/(1+.0157*CN)^20)*U-.065683* EXP
(-8.7513*U)
9802 LET P2=.3362*(1- EXP (-.03442*ER))
9804 LET P3=.0363* EXP (-4.6*U)*(1- EXP (-(CN/38.7)^4.97))
9806 LET P4=1+2.751*(1- EXP (-(ER/15.916)^8))
9808 LET P5=.334* EXP (-3.3*(ER/15)^3)+.746
9810 LET P6=P5* EXP (-(CN/18)^.368)
9812 LET P7=1+4.069*P6*G^.479* EXP (-1.347*G^.595-.17*G^2.5)
9814 LET P8=.7168*(1+1.076/(1+.0576*(ER-1)))
9816 LET P9=P8-.7913*(1- EXP (-(CN/20)^1.424))* ATN (2.481*(ER/
8)^.946)
9818 LET P10=.242*(ER-1)^.55
9820 LET P11=.6366*( EXP (-.3401*CN)-1)* ATN (1.263*(U/3)^1.629
)
9822 LET P12=P9+(1-P9)/(1+1.183*U^1.376)
9824 LET P13=1.695*P10/(.414+1.605*P10)
9826 LET P14=.8929+.1072*(1- EXP (-.42*(CN/20)^3.215))
9828 LET P15= ABS (1-.8928*(1+P11)*P12* EXP (-P13*G^1.092)/P14)
9830 LET FEF=P1*P2*((P3*P4+.1844*P7)*CN)^1.5763
9832 LET FOF=P1*P2*((P3*P4+.1844)*CN*P15)^1.5763
9834 LET EFEF=ER-(ER-EFEO)/(1+FEF)
9836 LET EFOF=ER-(ER-EFOO)/(1+FOF)
9838 LET Q1=.8695*U^.194
9840 LET Q2=1+.7519*G+.189*G^2.31
9842 LET Q3=.1975+(16.6+(8.399999/G)^6)^(-.387)+1/241* LOG(G^10
/(1+(G/3.4)^10))
```

```
9844 LET Q4=2*Q1/(Q2*( EXP (-G)*U^Q3+(2- EXP (-G))*U^(-Q3)))
9846 LET Q5=1.794+1.14* LOG(1+.638/(G+.5170001*G^2.43))
9848 LET Q6=.2305+ LOG(G^10/(1+(G/5.8)^10))/281.3+1/5.1* LOG(1+
.598*G^1.154)
9850 LET Q7=(10+190*G*G)/(1+82.3*G*G*G)
9852 LET Q8= EXP (-6.5-.95* LOG(G)-(G/.15)^5)
9854 LET Q9= LOG(Q7)*(Q8+1/16.5)
9856 LET Q10=(Q2*Q4-Q5* EXP ( LOG(U)*Q6*U^(-Q9)))/Q2
9858 LET FU=6+(2* PI -6)* EXP (-(30.666/U)^.7528)
9860 LET ZL=60/ SQR (EF)* LOG(FU/U+ SQR (1+(2/U)^2))
9862 LET ZE=ZL* SQR (EF/EFEO)/(1-(ZL/377)* SQR (EF)*Q4)-ZCE
9864 LET ZO=ZL* SQR (EF/EFOO)/(1-(ZL/377)* SQR (EF)*Q10)-ZCO
9866 RETURN
9868 REM Chain subroutine
9870 PRINT
9872 INPUT"Enter 'CONT':",N$
9874 IF N$="CONT" THEN CLS:CHAIN "MENU",10
9876 GOTO 9870
```

REFERENCES

[1] King, R.W.P., *Transmission-Line Theory,* New York: McGraw-Hill, 1955.

[2] Gunston, M.A.R., *Microwave Transmission-Line Impedance Data,* London: Van Nostrand Reinhold, 1972.

[3] Ortega, J.M., and W.C. Rheinboldt, *Iterative Solution of Nonlinear Equations in Several Variables,* New York: Academic Press, 1970.

[4] Himmelblau, D.M., *Applied Nonlinear Programming,* New York: McGraw-Hill, 1972.

[5] Wheeler, H.A., "The Transmission-Line Properties of a Round Wire Between Parallel Planes," *IRE Trans., Antennas and Propagation,* AP-3, No. 10, October 1955, pp. 203–207.

[6] Gupta, K.C., R. Garg, and R. Chadha, *Computer-Aided Design of Microwave Circuits,* Norwood, MA: Artech House, 1981.

[7] Shibata, H., Y. Kikuchi, and R. Terakado, "Characteristic Impedance of the Slab Line With an Anisotropic Dielectric," *IEEE Trans., Microwave Theory and Techniques,* MTT-33, No. 5, May 1985, pp. 419–421.

[8] Oberhetinger, F., und W. Magnus, *Anwendung der Elliptischen Funktionen in Physik und Technik,* Berlin: Springer-Verlag, 1949.

[9] Cohn, S.B., "Characteristic Impedance of Shielded-Strip Transmission Line," *IRE Trans., Microwave Theory and Techniques,* MTT-2, No. 7, July 1954, pp. 52–55.

[10] Rosłoniec, S., *Algorithms for Computer-Aided Design of Some Linear Microwave Circuits* (in Polish), Warsaw: Wydawnictwa Komunikacji i Łączności, 1987.

[11] Wheeler, H.A., "Transmission Line Properties of a Stripline Between Parallel Planes," *IEEE Trans., Microwave Theory and Techniques,* MTT-26, No. 11, November 1978, pp. 866–876.

[12] Atwater, H.A., "Tests of Microstrip Dispersion Formulas," *IEEE Trans., Microwave Theory and Techniques,* MTT-36, No. 3, March 1988, pp. 619–621.

[13] Hammerstad, E., and O. Jensen, "Accurate Models for Microstrip Computer-Aided Design," *IEEE MTT-S, International Symposium Digest,* Washington, DC, May 1980, pp. 407–409.

[14] Kirsching, M., and R.H. Jansen, "Accurate Model for Effective Dielectric Constant of Microstrip with Validity up to Millimeter Wave Frequencies," *Electronics Letters,* Vol. 18, No. 6, March 1982, pp. 272–273.

[15] Pramanick, P., and P. Bhartia, "Computer-Aided Design Models for Millimeter Wave Finlines and Suspended Substrate Lines," *IEEE Trans., Microwave Theory and Techniques,* MTT-33, No. 12, December 1985, pp. 1429–1435.

[16] Tomar, K.S., and P. Bhartia, "New Quasi-Static Models for Computer-Aided Design of Suspended and Inverted Microstrip Lines," *IEEE Trans., Microwave Theory and Techniques,* MTT-35, No. 4, April 1987, pp. 453–457.

[17] Cristal, E.G., "Coupled Circular Cylindrical Rods Between Parallel Ground Planes," *IEEE Trans., Microwave Theory and Techniques,* MTT-12, No. 7, July 1964, pp. 428–438.

[18] Rosłoniec, S., "An Improved Algorithm for the Computer-Aided Design of Coupled Slab Lines," *IEEE Trans., Microwave Theory and Techniques,* MTT-37, No. 1, January 1989, pp. 258–261.

[19] Stracca, G.B., G. Macchiarella, and M. Politi, "Numerical Analysis of Various Configurations of Slab Lines," *IEEE Trans., Microwave Theory and Techniques,* MTT-34, No. 3, March 1986, pp. 359–363.

[20] Cohn, S.B., "Shielded Coupled-Strip Transmission Lines," *IRE Trans., Microwave Theory and Techniques,* MTT-3, No. 10, October 1955, pp. 29–38.

[21] Edwards, T.C., Foundations of Microstrip Circuit Design, New York: John Wiley & Sons, 1982.

[22] Kirsching, M., and R.H. Jansen, "Accurate Wide-Range Design Equations for the Frequency Dependent Characteristic of Parallel Coupled Microstrip Lines," *IEEE Trans., Microwave Theory and Techniques,* MTT-32, No. 1, January 1984, pp. 83–90, also *IEEE Trans., Microwave Theory and Techniques,* MTT-33, No. 3, March 1985, p. 288.

[23] Rosłoniec, S., "Design of Coupled Microstrip Lines by Optimization Methods," *IEEE Trans., Microwave Theory and Techniques,* MTT-35, No. 11, November 1987, pp. 1072–1074.

[24] Gupta, K.C., R. Garg and I.J. Bahl, *Microstrip Lines and Slotlines,* Norwood, MA: Artech House, 1979.

[25] Hoffmann, R.K., *Handbook of Microwave Integrated Circuits,* Norwood, MA: Artech House, 1987.

APPENDIX

A.1 COMPLETE ELLIPTIC INTEGRALS OF THE FIRST KIND

The characteristic impedances of single and coupled triplate striplines with zero-thickness center conductors ($t \ll b$; see, e.g., Figure 6.3) are expressed by complete elliptic integrals of the first kind. Therefore, we give a short description of them together with a simple computational algorithm. In the mathematical literature (see, e.g., [1–3]), the complete elliptic integrals of the first kind are usually denoted by $K(k)$ and $K'(k)$. The integral $K(k)$ is defined as:

$$K(k) = \int_0^1 \frac{\mathrm{d}x}{\sqrt{(1 - x^2)(1 - k^2x^2)}} \tag{A.1}$$

where k $(0 \leqslant k \leqslant 1)$ is the modulus of $K(k)$. The associated (complementary) integral $K'(k)$ is defined as:

$$K'(k) = K(k') \tag{A.2}$$

where

$$k' = \sqrt{1 - k^2}\,.$$

The calculation of integral $K(k)$, when its modulus k is known, is not complicated, and can be done by using the algorithm presented below. Hence, let us consider two infinite mathematical series (a_n) and (b_n) defined as:

$$\begin{aligned} a_0 &= 1 + k & b_0 &= 1 - k \\ a_1 &= (a_0 + b_0)/2 & b_1 &= \sqrt{a_0 b_0} \\ a_2 &= (a_1 + b_1)/2 & b_2 &= \sqrt{a_1 b_1} \end{aligned} \tag{A.3}$$

$$\ldots \qquad \ldots$$

$$a_{n+1} = (a_n + b_n)/2 \quad b_{n+1} = \sqrt{a_n b_n}$$

Series defined in this manner converge to a common limit, usually denoted as $\mu(k)$. Then:

$$\lim_{n \to \infty} (a_n) = \lim_{n \to \infty} (b_n) = \mu(k) \tag{A.4}$$

Finally, the value of integral $K(k)$ is related to $\mu(k)$ as follows:

$$K(k) = \pi/[2\mu(k)] \tag{A.5}$$

The integral $K'(k)$ can be calculated in a similar way. Of course, in this case, we must calculate the limit $\mu(k')$ instead of $\mu(k)$.

Series (a_n) and (b_n) are rapidly convergent, and in the most practical cases, only a few iterations, for instance $n = 5$, must be taken into account.

Formulas (A.3) to (A.5) have been used as a mathematical basis of the computer procedure $K(k)$ presented below:

Listing of K(k)

```
10 REM K(k)
20 CLS
30 DEFDBL A-Z
40 PRINT "ELLIPTIC INTEGRAL K(k)"
50 PRINT
60 INPUT "k=",K
70 PRINT
80 LET PI=3.141592653#
90 IF K<0 THEN  PRINT "Error:k<0": GOTO 50
100 IF K >= 1 THEN  PRINT "Error:k >= 1": GOTO 50
110 LET A=1+K
120 LET B=1-K
130 LET C=(A+B)/2
140 LET D= SQR (A*B)
150 IF  ABS (C-D) <= 9.999999E-13 THEN  GOTO 190
160 LET A=C
170 LET B=D
180 GOTO 130
190 PRINT USING"K(k)=##.##########"; PI /(C+D)
200 STOP
```

We can test this procedure using the numerical results presented in Table A.1.

Table A.1
Some Calculation Results for the Complete Elliptic Integrals of the First Kind

k^2	$K(k)$	$K'(k)$
0.00	1.570 796 326 794	$\to \infty$
0.01	1.574 745 561 517	3.695 637 362 989
0.10	1.612 441 348 720	2.578 092 113 348
0.50	1.854 074 677 301	1.854 074 677 301
0.90	2.578 092 113 348	1.612 441 348 720
0.99	3.695 637 362 989	1.574 745 561 517
1.00	$\to \infty$	1.570 796 326 794

In transmission-line applications (see, e.g., Sections 6.4 and 6.8), the integrals under consideration are often encountered in the form of the ratio $(K(k)/K'(k)$. This ratio can be calculated directly from the following formula:

$$K(k)/K'(k) = \prod_{n=0}^{\infty} [(1 + k_n)/(1 + k'_n)] \tag{A.6}$$

where:

$$\begin{aligned}
k_0 &= k & k'_0 &= \sqrt{1 - k^2} \\
k_1 &= 2\sqrt{k_0}/(1 + k_0) & k'_1 &= 2\sqrt{k'_0}/(1 + k'_0) \\
k_2 &= 2\sqrt{k_1}/(1 + k_1) & k'_2 &= 2\sqrt{k'_1}/(1 + k'_1) \\
&\cdots & &\cdots \\
k_{n+1} &= 2\sqrt{k_n}/(1 + k_n) & k'_{n+1} &= 2\sqrt{k'_n}/(1 + k'_n)
\end{aligned}$$

We see that, in a computational process, only a limited number of factors of the product (A.6) can be taken into account. From the performed numerical analysis, we see that for $0.01 \leqslant k^2 \leqslant 0.99$, (A.6) ensures a very good accuracy of approximation if $n \geqslant 6$.

A.2 CHEBYSCHEV POLYNOMIALS OF THE FIRST KIND

From the approximation theory, it follows that the Chebyschev polynomials of the first kind, usually denoted as $T_n(x)$, allow us to find an approximation of an arbitrary function which reduces the maximum approximation error to minimum or to nearly minimum. They are generated from the sequence of cosine functions (i.e., cos(θ), cos(2θ), . . . , cos(nθ)), by using the transformation [4]:

$$\theta = \cos^{-1}(x) \tag{A.7}$$

Hence $T_0(x) = \cos(0) = 1$. On introducing (A.7), the first cosine function is transformed into the Chebyschev polynomial of degree one, namely:

$$T_1(x) = \cos(\theta) = \cos[\cos^{-1}(x)] = x \tag{A.8}$$

Similarly, the polynomial of degree two results from the trigonometric identity $\cos(2\theta) = 2\cos^2(\theta) - 1$. Consequently,

$$\cos(2\theta) = \cos[2\cos^{-1}(x)] = 2\cos^2[\cos^{-1}(x)] - 1$$

and

$$T_2(x) = 2x^2 - 1 \tag{A.9}$$

In general, the repeated application of the trigonometric identity

$$\cos(n\theta) = 2\cos(\theta)\cos[(n-1)\theta] - \cos[(n-2)\theta]$$

can be used to compute the high-order polynomials, and yields the recursion relation:

$$T_n(x) = 2xT_{n-1}(x) - T_{n-2}(x) \text{ for } n \geqslant 2 \tag{A.10}$$

Finally,

$$\begin{aligned} T_0(x) &= 1 \\ T_1(x) &= x \\ T_2(x) &= 2x^2 - 1 \\ T_3(x) &= 4x^3 - 3x \\ &\ldots \end{aligned}$$

The polynomials $T_i(x)$ of degrees 0 to 3 are shown in Figure A.1. Note that the Chebyschev polynomials (because of their $\cos(n\theta)$ origin) have $n + 1$ extreme values of equal magnitude (1) alternately positive and negative on the interval $[-1, 1]$.

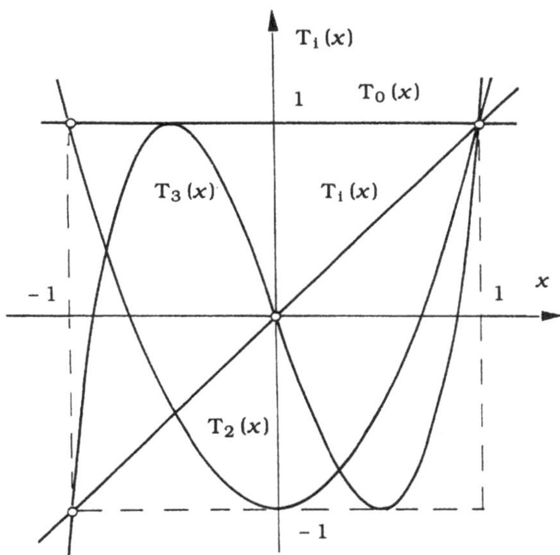

Figure A.1 The Chebyschev polynomials of the first kind of degrees 0, 1, 2 and 3.

A.3 SOME MATHEMATICAL AND PHYSICAL CONSTANTS

The SI system (International System of Units) is used throughout the book. The fundamental units are: meters (m) for length, kilograms (kg) for mass, seconds (s) for time, amperes (A) for electric current, kelvins (K) for temperature, and candelas (cd) for brightness. The supplementary units are: radians (rad) for plane angles, and steradians (sr) for space angles. All other units are derived.

The mathematical and physical constants presented below are expressed in SI units.

Mathematical Constants

$e = 2.718281 \cdots$ base for natural logarithms

$\pi = 3.141592 \cdots$

$j = \sqrt{-1}$ imaginary unit

Physical Constants

$c_0 = 2.997924\ 10^8$ m/s velocity of light in free space
$\epsilon_0 = 8.854184\ 10^{-12}$ F/m permittivity of free space
$\mu_0 = 4\pi\ 10^{-7}$ H/m permeability of free space
$\eta_0 = 120\pi$ ohm characteristic impedance of free space

The above physical constants are interrelated as follows:

$$c_0 = 1/\sqrt{\epsilon_0 \mu_0}$$
$$\eta_0 = \sqrt{\mu_0/\epsilon_0}$$

In the English technical literature, inches and mils are often used as units of length. These units are related to the meter as:

$$1 \text{ inch} = 25.4\ 10^{-3} \text{ m}$$
$$1 \text{ mil} = 0.001 \text{ inch} = 25.4\ \mu\text{m}$$

where μm denotes 0.000001th part of a meter.

REFERENCES

[1] King, L.V., *On the Direct Numerical Calculation of Elliptic Functions and Integrals,* Cambridge: Cambridge University Press, 1924.

[2] Oberhetinger, F. und W. Magnus, *Anwendung der elliptischen Funktionen in Physik und Technik,* Berlin: Springer-Verlag, 1949.

[3] Abramowitz, M., and I.A. Stegun, *Handbook of Mathematical Functions,* New York: Dover, 1954.

[4] Carnaham, B., H.A. Luther, and J.O. Wilkes, *Applied Numerical Methods,* New York: John Wiley & Sons, 1969.

INDEX

The Author

Stanislaw Rosloniec earned his MSc and PhD from the Warsaw Technical Institute (WTI) and has held several positions there over the past 18 years. He started in 1972 as a research and teaching assistant, and in 1976 he began his current position as an assistant professor in WTI's Institute of Radioelectronics. The author of 2 books and over 30 papers in the field of analysis and design of electrical circuits with distributed parameters, Dr. Rosloniec's main areas of interest are electrodynamics, microwave theory and techniques, and computer aided design of microwave circuits.

The Artech House Microwave Library

Analysis, Design, and Applications of Fin Lines, Bharathi Bhat and Shiban K. Koul

Capacitance, Inductance, and Crosstalk Analysis, Charles S. Walker

Computer-Aided Design of Microwave Circuits, K.C. Gupta, R. Garg, and R. Chadha

Design of Impedance-Matching Networks for RF and Microwave Amplifiers, Pieter L.D. Abrie

Digital Microwave Receivers, James B. Tsui

Electric Filters, Martin Hasler and Jacques Neirynck

E-Plane Integrated Circuits, P. Bhartia and P. Pramanick, eds.

Evanescent Mode Microwave Components, George Craven and Richard Skedd

Feedback Maximization, Boris J. Lurie

Filters with Helical and Folded Helical Resonators, Peter Vizmuller

GaAs FET Principles and Technology, J.V. DiLorenzo and D.D. Khandelwal, eds.

GaAs MESFET Circuit Design, Robert A. Soares, ed.

Handbook of Microwave Integrated Circuits, Reinmut K. Hoffmann

Handbook for the Mechanical Tolerancing of Waveguide Components, W.B.W. Alison

Handbook of Microwave Testing, Thomas S. Laverghetta

High Power Microwave Sources, Victor Granatstein and Igor Alexeff, eds.

Introduction to Microwaves, Fred E. Gardiol

LOSLIN: Lossy Line Calculation Software and User's Manual, Fred E. Gardiol

Lossy Transmission Lines, Fred E. Gardiol

Materials Handbook for Hybrid Microelectronics, J.A. King, ed.

MIC and MMIC Amplifier and Oscillator Circuit Design, Allen Sweet

Microelectronic Reliability, Volume I: Reliability, Test, and Diagnostics, Edward B. Hakim, ed.

Microelectronic Reliability, Volume II: Integrity Assessment and Assurance, Emiliano Pollino, ed.
Microstrip Antenna Design, K.C. Gupta and A. Benalla, eds.
Microstrip Lines and Slotlines, K.C. Gupta, R. Garg, and I.J. Bahl
Microwave Circulator Design, Douglas K. Linkhart
Microwave Engineers' Handbook: 2 volume set, Theodore Saad, ed.
Microwave Filters, Impedance Matching Networks, and Coupling Structures, G.L. Matthaei, L. Young, and E.M.T. Jones
Microwave Integrated Circuits, Jeffrey Frey and Kul Bhasin, eds.
Microwaves Made Simple: Principles and Applications, Stephen W. Cheung, Frederick H. Levien, *et al.*
Microwave and Millimeter Wave Heterostructure Transistors and Applications, F. Ali, ed.
Microwave Mixers, Stephen A. Maas
Microwave Transmission Design Data, Theodore Moreno
Microwave Transition Design, Jamal S. Izadian and Shahin M. Izadian
Microwave Transmission Line Filters, J.A.G. Malherbe
Microwave Transmission Line Couplers, J.A.G. Malherbe
Microwave Tubes, A.S. Gilmour, Jr.
MMIC Design: GaAs FETs and HEMTs, Peter H. Ladbrooke
Modern GaAs Processing Techniques, Ralph Williams
Modern Microwave Measurements and Techniques, Thomas S. Laverghetta
Monolithic Microwave Integrated Circuits: Technology and Design, Ravender Goyal, *et al.*
Nonlinear Microwave Circuits, Stephen A. Maas
Optical Control of Microwave Devices, Rainee N. Simons
Receiving Systems Design, Stephen J. Erst
Solid-State Microwave Devices, Thomas S. Laverghetta
Stripline Circuit Design, Harlan Howe, Jr.
Terrestrial Digital Microwave Communications, Ferdo Ivanek, *et al.*

www.ingramcontent.com/pod-product-compliance
Lightning Source LLC
Chambersburg PA
CBHW021430180326
41458CB00001B/205
9780890063545